A NATURALIST'S GUIDE TO THE
REPTILES
OF
INDIA
Bangladesh, Bhutan, Nepal, Pakistan and Sri Lanka

A NATURALIST'S GUIDE TO THE
REPTILES OF INDIA
Bangladesh, Bhutan, Nepal, Pakistan and Sri Lanka

Indraneil Das and Abhijit Das

Prakash Books

Reprinted in 2025

This edition of A Naturalist's Guide to the Reptiles of India is published and distributed in India by Prakash Books India Pvt. Ltd., 113 A Ansari Road, Daryaganj, New Delhi-110002, India, by arrangement with John Beaufoy Publishing Ltd.

10 9 8 7 6 5 4

Copyright © 2017 John Beaufoy Publishing Limited
Copyright in text © 2017 Indraneil Das and Abhijit Das
Copyright in photographs © 2017 in photographs: as listed below
Copyright in maps © 2017 John Beaufoy Publishing Limited

Photo credits:
Front cover: *main image* King Cobra (Manoj Nair); bottom row, left to right: Tokay Gecko (Abhijit Das), Indian Flapshell Turtle (Abhijit Das), Indian Chameleon (Indraneil Das); Back cover: Indian Krait (Abhijit Das); **Title page:** Maria's Lizard (Abhijit Das); **Contents page:** River Terrapin (Indraneil Das).
Main descriptions: photos are denoted by page number, followed by t (top), m (middle), b (bottom), l (left), c (centre) or r (right).
Steven C. Anderson 40t, **Rajeev Basumatary** 19br. **Ashok Captain/The Lisus** 92b, 105b, 118b, 125t. **Abhijit Das** 14b, 18t, 18b, 21t, 21b, 24tl, 25t, 25b, 30b, 31t, 32t, 32b, 33bl, 34t, 36b, 37t, 37b, 38t, 41tl, 41tr, 54tl, 56b, 57t, 59b, 60b, 63tc, 63tr, 64t, 65t, 66, 69t, 69b, 70tl, 70tr, 70b, 79t, 79b, 80t, 80b, 87tr, 90tr, 91bl, 91br, 92t, 93bl, 93br, 94bl, 94br, 95b, 96t, 96bl, 96br, 99br, 100tl, 100tr, 101b, 102t, 102b, 103r, 104tl, 104tr, 105t, 106b, 107bl, 107br, 108b, 109b, 110b, 111b, 112bl, 112br, 113t, 113b, 115b, 116t, 117tl, 117tr, 117b, 118t, 119t, 119b, 120r, 121t, 122r, 125t, 126r, 126b, 127t, 127bl, 128b, 129tl, 129tr, 132bl, 132br, 133, 137b, 141t, 142b, 144r, 145b, 146t, 149b, 150tl, 150tr, 150bl, 150br, 152b, 155t, 156b. **Indraneil Das** 137, 13b, 14t, 15, 16t, 16bl, 16br, 17tl, 16tr, 16bl, 16br, 19tl, 19tr, 19bl, 20r, 20b, 22t, 22b, 23t, 23b, 24tr, 24bl, 24br, 26t, 26b, 27tl, 27bl, 27br, 28, 29tl, 29tr, 29b, 30t, 31bl, 31br, 33t, 33br, 34b, 35t, 35b, 36tl, 36tr, 38bl, 38br, 39t, 39b, 40b, 41bl, 41bc, 41br, 42t, 42bl, 42br, 43t, 43b, 44bl, 44br, 45t, 45bl, 45br, 46t, 46b, 47t, 47b, 48t, 48b, 49t, 49b, 50t, 50b, 51t, 51b, 52tl, 52tr, 52bl, 52br, 53t, 53b, 54tr, 54b, 55t, 55b, 56tl, 56tr, 57b, 58t, 58b, 59r, 60t, 61t, 61b, 62t, 62b, 63tl, 63b, 64bl, 64br, 65b, 67t, 67b, 68t, 68b, 71m, 71bl, 71bc, 71br, 72t, 72b, 73t, 74t, 74b, 75t, 75b, 76t, 76b, 77t, 77b, 78t, 78b, 81t, 81b, 82tl, 82tr, 82bl, 82br, 83t, 83b, 84t, 84b, 85t, 85b, 86t, 86b, 87tl, 87b, 88t, 88b, 89t, 90l, 90r, 91tl, 91bc, 93t, 94tl, 94tr, 97t, 97b, 98t, 98b, 99tl, 99tr, 99bl, 100b, 101t, 102t, 104b, 107t, 108t, 110t, 112a, 114t, 114b, 115t, 116br, 120b, 122b, 123tl, 123tr, 123b, 124t, 124b, 127br, 128tl, 128tr, 130t, 131b, 132t, 134t, 135b, 136b, 137tl, 137tr, 138tl, 138tr, 138bl, 138br, 140t, 140b, 141b, 142tl, 142tr, 143, 144bl, 144br, 145tl, 145tr, 146m, 146bl, 146br, 147tl, 147tr, 147bl, 148t, 148b, 149t, 151b, 152t, 153bl, 153br, 154t, 154b, 155l, 155c, 155r, 156tl, 156tr, 156b. **David Jones** 27tr. **Muhammad Sharief Khan** 44t, 89b, 111t, 130b, 147br. **H. T. Lalremsanga** 106t, 116bl, 151t. **Manoj Nair** 134t, 135t, 139t. **Nikolai Orlov** 139b. **Klaus-Dieter Schulz** 95t, 103b. **Saibal Sengupta** 130b. **Alexander Teynié** 109t. **Gernot Vogel** 153t. **Raju Vyas** 121b. **George R. Zug/US National Museum** 71t.

All rights reserved. No part of this publication may be reproduced, stored in a retrieval system or transmitted in any form or by any means, electronic, mechanical, photocopying, recording or otherwise, without the prior written permission of the publishers.
Great care has been taken to maintain the accuracy of the information contained in this work. However, neither the publishers nor the author can be held responsible for any consequences arising from the use of the information contained therein.
The presentation of material in this publication and the geographical designations employed do not imply the expression of any opinion whatsoever on the part of the Publisher concerning the legal status of any country, territory or area, or concerning the delimitation of its frontiers or boundaries.

ISBN 978-93-86538-02-4

Edited by Krystyna Mayer
Designed by Gulmohur Press, India

Printed and bound in Malaysia by Times Offset (M) Sdn. Bhd.

Dedication
We dedicate this volume to our respective families, friends and colleagues.

·CONTENTS·

Introduction 6

Climate and Vegetation 6

Conservation of Reptiles 9

Snake-bite Management 10

About This Book 11

Glossary 12

Species Descriptions and Photographs 13

Checklist of the Reptiles of South Asia 157

Further Reading 171

Index 173

INTRODUCTION

Known as the Indian subcontinent and the Indian Region, South Asia is one of tropical Asia's regions of the greatest biodiversity. The region includes the countries next to or in the proximity of the Indian Ocean, comprising Bangladesh, Bhutan, India, the Maldives, Nepal, Pakistan and Sri Lanka. The geographic barriers that impede faunal movement into or out of the area include oceans, mountains and floodplains, and justify the recognition of the region as a distinct biogeographic unit.

South Asia is exceptionally diverse in reptiles (more than 700 species). This richness is linked to the large size of the area, covering about 4.4 million square kilometres, and its location at the crossroads of two distinctive biogeographic realms, the Palaearctic and the Oriental. Consequently, there is much greater species diversity here than in even larger areas, such as China, North America north of the Rio Grande River, and Europe east of the Ural Mountains. The fauna of the Indian region is comparable to several tropical regions in the Indo-Pacific and Neotropical regions, such as Indonesia, Australia, Colombia and Brazil.

CLIMATE AND VEGETATION

The habitat range in south Asia includes coral reefs, mangrove swamps, closed-canopy rainforests, thorn-scrub vegetation and deserts. The region also includes several environments with physical extremes, such as the highest mountains and wettest locations on Earth. Human intervention has reduced both the extent and quality of natural areas worldwide, and south Asia has been particularly affected. Given below are the countries of the subcontinent and their major ecological regions.

Bangladesh

Bangladesh lies at between 21–27° N and 88–93° E, and is dominated by the floodplains of the Ganga and Brahmaputra Rivers. Mostly a low-lying, small (total land area 143,998km^2) country, hilly areas occur only in the north-east (Sylhet) and south-east (Chittagong Hill Tracts), accounting for 10 per cent of the land area. The climate is subtropical, with an annual rainfall of 1,500–5,000mm, most of which comes during the south-west monsoon. Bangladesh has one of the world's densest human populations (9,388/1,000ha), and just 11 per cent of the land area remains under forest cover. Forest types represented include open deciduous forests, which occur on the dry, exposed southern slopes of the Chittagong Hill Tracts, moist deciduous forests, dominated by the Sal (*Shorea robusta*), and found throughout the plains, and finally the tidal forests, largely mangroves, that may be seen in the Sundarbans, the delta of the Ganga and Brahmaputra Rivers.

Bhutan

Bhutan is situated between 88° 45' and 92° 10' E and 26° 40' and 28° 15' N in the eastern Himalayas. Its 38,394km^2 territory is largely mountainous, though the southern parts include the northern plains of the Brahmaputra River. Altitudinally, habitats range from 200m to more than 7,500m, and the climate ranges from hot and humid in the south, to alpine in the north. Bhutan therefore shows greater than expected ecological diversity for

CLIMATE & VEGETATION

its size. Forest cover comprises 68 per cent of the land area, and forestry plays a major part in the economy.

India

India is a country with exceptional reptile diversity. With a land area of 3,287,263km^2 and a variety of ecological conditions, it abuts the Himalayan range and extend southwards to the tip of the peninsula.

There are several major ecophysiographical regions in India.

The **Andaman and Nicobar archipelago** lies between 05° 40' N and 92° 10' E, in the Bay of Bengal. These islands form a chain of submarine mountains that sprawl in a crescent between Cape Negrais in Myanmar and Achin Head in Sumatra, Indonesia. The total land area of these islands is an estimated 8,293km^2. Average annual rainfall exceeds 3,000mm, and habitats represented include coral reefs, mangroves and rainforests on hills of up to 700m.

The **Deccan** is a flat plain comprising much of the Indian peninsula, excluding the hill ranges to the east and west, and south of the Himalayas. Until the Miocene and Pliocene Epochs (and perhaps as recently as the Late Pleistocene), evergreen forests were widespread here, and their transformation to deciduous forests was probably the result of the southern shift of the Equator, the uplift of the Himalayas and the rise of the Western Ghats, causing a reduction in rainfall, in addition to human activities over the past 10,000 years.

The **Eastern Ghats** represent a weathered relict of the peninsular plateau, marked by a series of low, isolated hills that run from the Khondmal in the Baudh-Khondmal region of Odisha State, southwards to central Tamil Nadu State, where they veer off towards the south-west to meet the Western Ghats in the Nilgiris. The northern and southern sections of the Eastern Ghats are separated by the Godavari delta, approximately 130km in width, with other important breaks including the Mahanadi and Krishna Rivers. The southern subzone is quite arid, with dry deciduous and thorn scrub, while the northern part is relatively moist, with both dry and moist deciduous forests.

The Himalayan mountain range includes some of the highest mountains on Earth. Several rivers, including the Ganga, Brahmaputra, Yangtze, Indus and Mekong, originate from this range. The Himalayas, including the Trans-Himalayas, cover an area of 236,300km^2, including parts of Pakistan, India, Nepal and Bhutan. Forest types represented range from moist deciduous, through subtropical broadleaved, to coniferous, mixed coniferous and alpine scrub forests, in addition to the *terai*, a swampy belt, the *bhabars*, which are deep, boulder deposits, and the *duns*, which are broad elevated valleys.

The **Eastern Himalayas** are wetter than the western part, receiving an annual average of at least 4,000mm of rainfall, and often much more. However, the winter months are relatively dry. At 4,000–5,000m are the alpine pastures, considered 'grasslands' though few grasses are represented, the dominant vegetation being perennial mesophytic herbs.

With rainfall exceeding 2,000mm, the **North-east** supports tropical and subtropical vegetation, including moist deciduous, semi-evergreen and temperate montane forests. Tropical evergreen forests in the region comprise forests with a three-tiered structure, the highest of which reach about 46m above the forest floor. Climatic fluctuations during the year are minimal, temperatures on average being 20–30° C in the plains.

Climate & Vegetation

The **North-west** includes the extreme western parts of India and the eastern districts of Pakistan, constituting the Western Himalayas, and is bounded by the Indus and Nara Valleys in the west, the Aravalis in the east and the Kachchh to the south. To the north lie the plains of the Sutlej and Chambal Rivers. The region is mostly composed of hills, stony plateaus and peneplains. Severe winters characterize the zone, which is outside the influence of the monsoons. Rainfall is 250–500mm annually, and the mean maximum temperature is more than 45° C. Thorny thickets are the common woody vegetation, and in the western parts, in the Nara region of Pakistan, the vegetation is sparse, consisting of xerophytic shrubs.

Zanskar, Ladakh and Karakorum dominate the landscape of the Trans-Himalayas (outer Himalayas). To the east, Zanskar and Ladakh reach down to the Tibetan plateau, where the region is marked with brackish marshes and bogs. The region is composed of mountains that are up to 6,600m high, and sandy valleys drained by the Indus. The dry landscape is due in part to the extreme low temperatures (below 0° C) that inhibit the absorption of water by roots of plants during the winter and early spring when occasional showers take place. The vegetation of the Trans-Himalayas includes coniferous forests as well as alpine steppe. In general, the rainfall increases along a west–east gradient, reaching 1,000m in the Kumaon region. At higher altitudes, the vegetation is xerophytic.

The **Western Ghats** run along the west coast of peninsular India 50–100km inland, and are a series of hill ranges often isolated from each other by low-lying savannahs. The hill ranges of the Nilgiris, Anaimalais and Palnis are the highest, reaching 450–1,500m, and receive average annual rainfall in excess of 2,000mm.

The Maldives

This archipelago lies between 08° N to 01° S and 72–74° E, and extends 756km south-west of peninsular India. The archipelago has a land area of 298km^2, and is composed of interrupted double chains of 26 coral atolls, resting on a submerged mountain range. There are 1,192 islands; 202 are inhabited, and all are small (average size 0.7km^2, the largest 13km^2), low lying and have average elevations of 1.5–2m. These islands are influenced by the two monsoons, the south-west in April–August and the north-east in October–February. Rainfall is evenly distributed throughout the year, averaging 1,950mm per annum. There is little seasonal variation in the temperature of between 24° C and 30° C. The island surface consists of coral sand with no topsoil, resulting in scrub vegetation. A few islands have large evergreen trees that form small, jungle-like growth. Small patches of pioneering mangrove species colonize the seaward side of these marshes on the inner edge of the lagoonal beach.

Nepal

Situated between 80–88° E and 26–30° 5' N, Nepal is between the Himalayas and the Indian peninsula, and covers a land area of 147,181km^2, which includes a vegetational and altitudinal range from lowland subtropical forests to the alpine forests associated with the highest mountain peak on Earth. Nepal consists of parallel physiographic zones from south to north that run the length of the country. The southernmost zone (elevation 80–300m) is the *terai*, an alluvial plain composed of sands, silts and clays. To its north lie the Churia Hills (average elevation 1,200m), which are primarily sedimentary rocks composed of sand, shale and gravels derived

Conservation of Reptiles

from weathering of the northern ranges. Further north is the Mahabharat Lekh (elevation 600–4,500m), composed of metamorphic and igneous rocks. Within the range lie the Midlands, a large area of fertile valleys and ridges in central Nepal. Further north lies the Himalayan Range (average elevation 6,100m), which includes some of the highest mountain peaks in the world.

Pakistan

Pakistan stretches between 23° 4' N to 36° 55' N and 60° 52' E to 75° 23' E, and is the westernmost country of south Asia, covering a land area of 347,190km^2. The country stretches from the arid cliffs of the Arabian sea coast to the permanent snow fields of the Pamir Wakhan. Biogeographically, the two components of Pakistan are the North-west and the Trans-Himalayas. The former zone includes extreme western India, the areas of Pakistan included being the Indus and Nara Valleys. The region is composed primarily of hills, and stony plateaus or peneplains. The rainfall is 250–500mm a year, and the mean maximum temperature is more than 45° C. Thorny thickets are the common vegetation to be seen here. Annual rainfall in the desert peneplane is 250mm or less. In the western parts such as in the Nara region of Pakistan, the vegetation is sparse, consisting of xerophytic shrub-like forms.

The Zanskar and Ladakh Rivers reach down to the Tibetan plateau, where the region is marked by brackish marshes and bogs. The outer Himalayas is composed of mountains that are up to 6,600m high, and sandy valleys drained by the Indus River. The dry climate is due in part to the extreme low temperatures (below 0° C). The vegetation includes subtropical evergreen, coniferous forests and alpine steppe.

Sri Lanka

The 65,610km^2 continental island of Sri Lanka lies between 05° 55'–09° 51' N and 79° 41'–81° 54' E, and is generally divided into a dry zone, including the northern half and the east coast (65 per cent of land area); a wet zone including the south-west (23 per cent); and an intermediate zone, including the centre of the southern half of the island (12 per cent). The wet zone of Sri Lanka is the only climatically aseasonal area between Malesia and the eastern coast of Madagascar. Sri Lanka's connection to the mainland, for the first time during the Miocene Epoch and many times subsequently, has lead to the invasion of many distinctly Indian taxa, although endemicity in the herpetofauna is high. The temperature range is 15–24° C in the southern highlands to 27–35° C in the northern and eastern plains. The wet zone, where most of the region's biodiversity is concentrated, is also the area that supports the greatest human population density (approximately two-thirds of the island's people).

Conservation of Reptiles

The Indian subcontinent harbours a rich reptile diversity. This follows from its long isolation from other land masses, the variety of tropical and temperate habitats it contains, and a network of protected areas. Two of the world's global hotspots of biodiversity fall within the region – the Eastern Himalayas (including north-east India, Nepal and Bhutan) and the Sri Lankan-Western Ghats hotspot. The biota of the region's dry zones show distinctive lineages, several of which are regional endemics.

South Asia is also home to some of the world's highest human densities, and as may be expected, anthropogenic pressures on natural habitats is immense. Species adapted to forested habitats are increasingly under pressure. Aquatic reptiles, on the other hand, are threatened by a number of factors, such as excessive fishing, pollution, and modification of habitats, including through the construction of dams, and mineral and petroleum extraction. Such activities have been linked to the destruction of Gharial- and turtle-nesting habitats. Shifting cultivation, as well as the establishment of monoculture plantation and encroachment, are some of the prime drivers of forest loss in this region.

Linear infrastructure projects, such as highways and railways, also pose an ever-increasing threat to wildlife, including that of reptiles. Such projects tend to bifurcate forested areas, including many that are protected, causing both habitat fragmentation and the removal of individuals as roadkills. Appropriate mitigational measures are thus urgently needed to enhance reptile conservation.

Several reptile species are persecuted out of fear and ignorance. During the wet season, many snakes encountered within human habitation are 'rescued' by well-meaning members of the public as well as by official governmental agencies (including local forest departments), but are unfortunately released in the nearest forest patches. The possible impact of such practices on snake populations, and on the local biodiversity, remains to be studied.

Turtles, tortoises, monitor lizards and snakes are subject to illegal trade in many parts of the region. Many are exploited for food and traditional medicine, especially to the east of the subcontinent, and most large-growing turtles have declined to the point of commercial extinction. A few of the more brightly coloured species, such as the Indian Star Tortoise and some of the tent turtles, are in demand from the international pet trade, and are harvested in vast numbers.

Many, if not all, snakes, monitor lizards, pythons and turtles are protected by law in the countries of the region. Killing, capture or removal are thus prohibited by law. One of the cornerstones in the conservation of reptiles is the spread of a protected areas network across the region, especially in India, which covers many distinct habitat types, including mangrove areas, alluvial grassland, marine and riverine areas, tropical evergreen forest, dry and mixed deciduous forests, and deserts. One hopes that such sites would provide appropriate habitats for the subcontinent's reptile fauna for the benefit of future generations.

SNAKE-BITE MANAGEMENT

A number of people, especially in rural areas of India, Sri Lanka and Bangladesh, are treated for snake bite annually. Indeed, a number of deadly snake species inhabit the region, both on land and in coastal areas. However, a majority of mortality is from species adapted to environments modified by humans, such as fields of rice paddy and rural areas, where rodents are prevalent. The main snakes responsible for nearly 90 per cent of all bites in the region are the Spectacled Cobra, Monocled Cobra, Saw-scaled Viper, Common Krait and Russell's Viper. Geographical variation in venom composition and effects remain poorly known, and additional species of venomous snake may await scientific description.

ABOUT THIS BOOK

A majority of the region's snakes being non-venomous, and the chances of getting bitten being remote, it is important to be able to differentiate the venomous snakes from the harmless ones. Vipers (such as Russell's Viper) and 'green' pit vipers (like those in the genera *Parias*, *Popeia*, *Protobothrops* and *Trimeresurus*) are relatively slow-moving snakes, with narrow necks and enlarged heads, whose fangs can be folded when not in use. Cobras (*Naja* and *Ophiophagus*) are large, heavy-bodied snakes, with the ability to raise a hood; they have short, fixed fangs. Coral snakes (*Calliophis* and *Sinomicrurus*) and kraits (*Bungarus*) are close relatives of cobras, but cannot raise their hoods. Finally, sea snakes (including *Hydrophis* and *Laticauda*) are large, slender- or heavy-bodied snakes that live in the sea or in coastal areas (though a few can travel a few kilometres upriver). All sea snakes have short, fixed fangs.

Most non-venomous snakes (and all crocodilians, many lizards and turtles) can bite humans when cornered or inappropriately handled, which may lead to bleeding and secondary infection, and a bite from any large snake, be it a rat snake or a python, can be painful.

Provided below are a few dos and don'ts to follow when visiting places where venomous or unknown snakes have been sighted:

- Do not put your hands inside dark areas, such as cracks or holes that may shelter a snake.
- Wear shoes or ankle-length boots that conceal the lower foot, especially in areas with tall grass, where large vipers may shelter.
- Carry a reliable flashlight or wear a headlamp when moving in the dark.
- Keep snakes away from human residences by keeping your surroundings clean. Refuse attracts rats, which in turn attract rat-eating snakes such as cobras.
- Avoid attempting to kill or capture snakes (unless you have received appropriate training, in addition to holding permits/licenses from the relevant authorities).

Anti-venom sera is the only scientifically known cure for cases of envenomation by the region's medically important snakes. Most health centres stock anti-venom, especially in areas where snake bites are common. In the event of a venomous snake bite, the affected person needs to be kept calm and warm, and taken to a hospital as quickly as possible. The region around the bite should be immobilized with a stiff cloth bandage (*never* a tight torniquet). A description or photograph of the snake with a mobile phone will help medical staff provide appropriate treatment, as the neurotoxic venom of cobras, kraits and coral snakes acts differently from the haemotoxic venom of vipers. It is important that the area affected is not cut or sucked, as such measures are likely to complicate the treatment.

ABOUT THIS BOOK

This book discusses representative reptile species that an average visitor or a resident of India and the surrounding countries is likely to encounter. A number of the species are nonetheless rare, and are illustrated here for the first time in a printed work. The aim of this volume is to aid rapid field identification for anyone interested in snakes and other reptiles (such identification also being useful for biodiversity surveys, and necessary for conservation and management).

■ Glossary ■

Typical Snakes

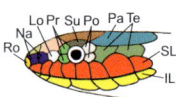
Lateral aspect of head

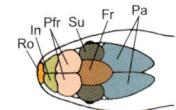
Dorsal aspect of head

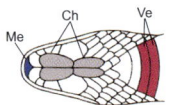
Ventral aspect of head

Skinks

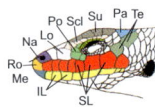
Lateral aspect of head

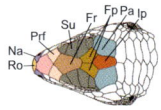
Dorsal aspect of head

Turtles, Terrapins & Tortoises

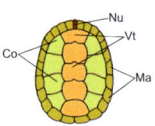
Carapace (Dorsal aspect)

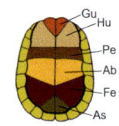
Plastron (Ventral aspect)

Major scales in reptiles. Top row, from left to right: snake head from side, top and bottom; bottom row, from left to right: skink head from side and top; turtle shell from top and bottom.

Key To Abbreviations

Ab abdominal scute, An anal plate, As anal scute, Ch chin shield, Cl cloaca/vent, Co costal, Fe femoral scute, Fp frontoparietal, Fr frontal, Gu gular scute, Hu humeral scute, IL infralabial, In internasal, Ip interparietal, Lo loreal, Ma marginal scute, Me mental, Na nasal, Nu nuchal, Pa parietal, Pe pectoral scute, Po postocular, Pr preocular, Prf prefrontal, Ro rostral, Sc subcaudal, Scl supraciliary, SL supralabial, Su supraocular, Te temporal, Ve ventral, Vt vertebral scute.

Glossary

aquatic Living in water.
arboreal Living in trees or in other vegetation away from the ground.
canopy Layer of vegetation above the ground, usually comprising tree branches and epiphytes.
clutch Total number of eggs laid by a female at a time.
courtship Behaviour preceding mating.
crepuscular Active during dawn and dusk.
depressed Flattened from top to bottom.
diurnal Active during day.
dorsum Dorsal surface of body, excluding head and tail.
femoral pores Pores on femoral region of some geckos.
fossorial Living underground.
infralabial Scales on lower lip.
keel Narrow prominent ridge.
lamella (pl. lamellae) Pad under digit in lizards (also **scansor**).
middorsal scales Scales around middle of body.
nocturnal Active during night.
oviparous Laying eggs.
ovoviviparous Form of reproduction when the eggs develop within the body of the mother, which does not provide nutrition other than the yolk.
preanal pores Pores situated in front of cloaca in geckos.
prefrontals Paired scales on anterior margin of orbit of eye, usually bounded by the frontal.
recurved Curved or bent.
reticulated Arranged like a net.
scansor Pads under digits in geckos (also **lamellae**).
scute Horny epidermal shield.
serrated With a saw-toothed edge.
subcaudal Scales below tail.
supralabial Scales on upper lip.
tubercle Knot-like projection.
tympanum Ear-drum.
ventral Scales under body, from throat to vent.
vermiculation Pattern consisting of worm-like markings.
vertebral Pertaining to region of backbone.
viviparous Live bearing, whereby embryo obtains additional nourishment from mother, in addition to yolk.
zygodactylous Of feet; the toes of each foot are arranged in pairs opposing each other.

▪ LAND TORTOISES ▪

> **TESTUDINIDAE (LAND TORTOISES)**
> The family of true tortoises comprises 59 described species. Land tortoises are mostly herbivorous (but may scavenge on carrion). They have thick shells, columnar forelimbs, elephantine hindlimbs, fingers and toes without webbing, heavily scaled outer faces of forelimbs, and lack axillary and inguinal glands.

Indian Star Tortoise ▪ *Geochelone elegans* 38cm
(*Gujarati* Khad No Kachba, Suraj Kachba, Zamin No Kachabo; *Kannada* Kal Aame; *Marwari* Khar Kachebo; *Oriya* Bali Kaichha; *Rajasthani* Bhumi Kachba; *Sinhalese* Hooni-yan Ibba, Mal Ibba, Makaral Ibba, Taruka Ibba; *Tamil* Kal Amai, Kattu Amai; *Telugu* Meta Tabelu; *Urdu* Tariwala Kachua, Satara Kachoor)

DESCRIPTION Carapace elongated in adults, rounded in juveniles, domed dorsally; weak bicuspid or tricuspid upper jaw; several distinct conical tubercles on thigh; carapace and plastron star marked with pattern of dark brown or black on yellow or beige; superimposed dark colour especially prominent in juveniles. **DISTRIBUTION** North-western, south-eastern and southern India, eastern Pakistan, and northern and eastern Sri Lanka. **HABITAT AND HABITS** Found in scrub forests and edges of deserts, agricultural fields, teak forests, grassland and thorn scrub. Diet largely herbivorous; known to eat grass and flowers, and also scavenges on animal matter. Clutches comprise 1–10 eggs, 40 51 x 31–37mm. Incubation period 47–178 days.

Elongated Tortoise ▪ *Indotestudo elongata* 33cm
(*Assamese* Halodia Kasso; *Bengali* Bon Kochchop, Gecho Kochchop, Pahari Haldey Kochchop; *Hindi* Parbati, Suryamukhi; *Khasia* Hunro; *Nepali* Ageri, Bhain Kachua, Padini; *Oriya* Mati Kaichha)

DESCRIPTION Carapace domed, highest point in vertebral III, flattened dorsally with arching sides; posterior marginals slightly flared and serrated (especially in juveniles); shell broadest posteriorly; plastron elongated, with deep notch posteriorly; limbs heavily scaled, club-like, bearing 5 claws each; nuchal scute long and narrow; carapace and plastron yellowish-brown or olive, with scattered black blotches; plastron sometimes unpatterned. **DISTRIBUTION** Northern to north-eastern India, Nepal, Bangladesh and Bhutan; also Southeast Asia. **HABITAT AND HABITS** Found in deciduous and evergreen forests. Diet includes leaves, fallen flowers and fruits, as well as fungi and occasionally, through scavenging, dead animals and slugs. Clutches comprise 1–7 eggs, 50 x 37mm. Incubation period 96–165 days. Hatchlings 49mm.

■ LAND TORTOISES ■

Travancore Tortoise ■ *Indotestudo travancorica* 33cm
(*Kannada* Betta Aame, Gudde Aame, Kadu Aame; *Kanis* Kal Ama, Vengala Amai; *Katumaran* Kar Aamai, Valli Aamai; *Malayalam* Churel Ama, Churelkata Amah; *Tamil* Peria Amai)

DESCRIPTION Carapace elongated, somewhat depressed, broader posteriorly in adults; shell relatively more rounded in juveniles; nuchal absent; tail ends in claw-like spur; shell olive or brown with black blotches on each scute. **DISTRIBUTION** Central and southern Western Ghats of south-western India. **HABITAT AND HABITS** Found in moist evergreen and semi-evergreen forests to an altitude of at least 1,000m. Mostly herbivorous, feeding on grass, fungi, bamboo shoots, fallen fruits and flowers, and also on insects, frogs and carrion. Clutches comprise 1–5 eggs, 40–58 x 31–44mm. Incubation period 146–149 days. Hatchlings 55–60mm.

Asian Giant Tortoise ■ *Manouria emys* 50cm
(*Bengali* Pahari Kochchop; *Chakma* Mon Dur; *Khasia* Phrau)

DESCRIPTION Carapace relatively low and rounded; vertebral region depressed; distinct growth rings on scutes of carapace; outer surfaces of forelimbs have large scales; paired tuberculate scales on thighs; carapace blackish-brown; plastron lighter; limbs dark brown to greyish-brown. **DISTRIBUTION** North-eastern India and Bangladesh; also southern China and Southeast Asia. **HABITAT AND HABITS** Found in evergreen forests, especially with hill streams and dense leaf litter. Largely herbivorous, though insects and frogs also eaten. Constructs mound nest by sweeping leaf litter, and lays 23–51 hard-shelled eggs of diameter 51–54mm. Guards nest, attacking egg predators. Hatchlings take 60–75 days to emerge, and measure 60–66mm.

LAND TORTOISES/POND TURTLES

Central Asian Tortoise ■ *Testudo horsfieldii* 22cm
(*Baluchi* Lach Pusht, Sang Toti, Tosh Bakke; *Pushtu* Sang Pusht; *Urdu* Bari Kachua, Sham Shatu)

DESCRIPTION Carapace rounded and domed, with depressed vertebral region; foot has 4 toes; tail ends in claw-like structure; carapace pale brown to olive, with dark blotches; plastron has large dark blotches; head and limbs greyish-yellow. **DISTRIBUTION** Balochistan and North-West Frontier Province in Pakistan; range extends west to shores of Caspian Sea. **HABITAT AND HABITS** Found in loamy and rocky, arid habitats, from plains to about 1,200m, though more dense around oases and in vicinity of wetlands. Excavates burrows 80–2,000cm in length, presumably critical for hibernation during winters and for avoiding the midday sun. Diet includes leaves, flowers and fruits. Clutches comprise 1–5 eggs, 42–43 x 30mm. Incubation period 62–82 days. Hatchlings 32–34mm.

GEOEMYDIDAE (Pond Turtles)
This is one of the largest and most diverse families of turtles, with 70 living species that may be aquatic or terrestial, ranging in size from under 10cm to some that reach nearly a metre in shell length. Members of the family inhabit fresh waters, coastal areas and especially, lowland forests of the subtropics and tropics of Asia, Europe and North Africa, with a single genus in Central and South America. Morphological characters associated with the family include neck that is retracted vertically, carapace with 24 marginal scutes, lack of mesoplastron, and pectoral and abdominal scutes contacting marginal scutes. Eggs are hard shelled, elongated and buried along the banks of water bodies.

POND TURTLES

River Terrapin ■ *Batagur baska* 59cm
(*Bengali* Bala Katha; *Bali* Katha, Boro ketho, Pora Katha, Ram Kachim, Sona Katha, Sundhi)

DESCRIPTION Carapace domed, heavily buttressed; long plastron; head small with narrow, upturned snout; forehead covered with small scales; jaws serrated; 4 claws on each forelimb, which is has wide webbing; carapace olive-grey or brown, and head similar coloured but lighter on sides; plastron unpatterned yellow; breeding males develop black

forehead and back of neck; front portion of neck bright red. **DISTRIBUTION** Estuaries such as the Sunderbans, Bhitarkanika and mouth of Subarnarekha River of India and Bangladesh. Also Myanmar. **HABITAT AND HABITS** Occurs in mouths of large rivers with mangroves. Fruits of *Sonneratia* are staple dietary item; leaves, stems and fruits also consumed, besides molluscs, crustaceans and fish. Nests on sea coasts. Clutches comprise 10–30 eggs, 69–75 x 39–45mm. Incubation period 65–66 days.

Three-striped Roofed Turtle ■ *Batagur dhongoka* 48cm
(*Bengali* Boro Katha, Sada Katha; *Hindi* Dhona, Dhoor, Dhor, Dhundi; *Nepali* Dhodari, Dond Chhane Kachhuwa; *Tharu*: Puberi)

DESCRIPTION Carapace elevated, oval, flaring at back, with rounded rim; keel ends in knob on third vertebral; snout slightly upturned; upper jaw has weak notch; digits entirely webbed; carapace brownish-grey, olive or greyish-yellow, with black or dark brown stripes on vertebral and pleural regions, as well as marginal edges; plastron yellow or cream; head and neck greyish-cream, with cream or yellow stripe from snout, across eyes and tympanum, to back of head. **DISTRIBUTION** River Ganga and its tributaries, including the Chambal, in northern India, Nepal and Bangladesh.

HABITAT AND HABITS Found in large rivers. Males omnivorous, eating water plants and molluscs; females take vegetable matter. Clutches comprise 21–35 eggs, 52–66 x 32–41mm. Incubation period 56–89 days.

Female   *Male*

◾ POND TURTLES ◾

Painted Roofed Turtle ◾ *Batagur kachuga* 56cm
(*Bengali* Adi Kori Katha; *Hindi* Lal Tilakwala, Sal; *Nepali* Rangin Chhane Kachhuwa)

DESCRIPTION Carapace moderately domed, oval; posterior marginals of juveniles strongly serrated, forming 5–7 sharp spines; plastron narrow; snout slightly upturned; upper jaw weakly bicuspid; carapace brownish-olive in males, dark brown or black in females; plastron of both sexes cream or yellow; adult males have blue-black head, broad red patch from tip of snout to occiput, 2 yellow stripes on sides of head and 6 red stripes on cream-coloured neck; adult females have yellow or silvery mandibles and dark brownish-black heads. **DISTRIBUTION** River Ganga and its tributaries in India, Bangladesh and Nepal. **HABITAT AND HABITS** Found in large rivers. Basks on sand banks, rocks and logs. Herbivorous, feeding on water plants. Clutches comprise 11–30 eggs, 64–75 x 38–46mm.

Female

Male

Malayan Box Turtle ◾ *Cuora amboinensis* 21.6cm
(*Assamese* Jap Dura; *Bengali* Chapa Katha, Diba Kochchop; *Car Nicobarese* Penyut, Takurab, Ta-penyut; *Central Nicobarese* Uptepe; *Manipuri* Thanggu; *Mising* Kop-kadera; *South Nicobarese* Hetain/Itain)

DESCRIPTION Carapace high domed and smooth, with single keel in adults; juveniles have 2 additional keels; carapace olive, brown or nearly black; plastron yellow or cream, with single black blotch; face yellow striped. **DISTRIBUTION** North-eastern India and Nicobar Islands, Bangladesh and Bhutan; also Myanmar, Thailand, Indo-China, Indo-Malaya and the Philippines. **HABITAT AND HABITS** Found in lakes, small rivers, grassland, marshes, mangrove swamps and also agricultural areas. Primarily herbivorous, feeding on water plants and fungi; worms and aquatic insects also eaten. Clutches comprise 1–6 eggs, 40–55 x 25–34mm. Incubation period 45–90 days.

POND TURTLES

Keeled Box Turtle ■ *Cuora mouhotii* 18cm
(*Assamese* Siria Jap Dura)

DESCRIPTION Carapace elongated, flat topped, with 3 prominent keels; marginals serrated posteriorly, and sometimes anteriorly as well; weak hinge across plastron in adult females; upper jaw hooked; digits half webbed; tail extremely long in juveniles, relatively shorter in adults; carapace dark or light brown; plastron yellow or light brown, with dark brown blotches on each scute; iris red. DISTRIBUTION North-eastern India, Bangladesh and Bhutan; also Southeast Asia and eastern China. HABITAT AND HABITS Found in evergreen hill forests, and associated with leaf litter. Omnivorous. Clutches comprises 1–5 eggs, 40–56 x 25–27mm. Incubation period 90–101 days.

Gemel's Leaf Turtle ■ *Cyclemys gemeli* 25cm
(*Bengali* Sil Kathua; *Manipuri* Thanggu; *Nepali* Thateru)

DESCRIPTION Shell oval, depressed, bearing 3 keels; shell widest at marginal VIII area; posterior marginals weakly serrated; enlarged scales on forehead; plastron elongated, with hinge in adults at around 23cm shell length; carapace and plastron brown to olive-brown, with dark radiating lines; head brown to blackish-brown, without distinct stripes. DISTRIBUTION Northern to north-eastern India, Bangladesh, Bhutan and Nepal. HABITAT AND HABITS Found in streams, small rivers, and ponds in low hills and plains. Juveniles more aquatic than adults. Omnivorous, feeding on figs and invertebrates. Clutches comprise 2–4 eggs. Incubation period 75 days.

▪ Pond Turtles ▪

Spotted Pond Turtle ▪ *Geoclemys hamiltonii* 41cm
(*Assamese* Nal Dura; *Bengali* Bagh Kathua, Bhuna Kaitta, Bhut Katha, Kalo Katha, Mogom Kathua, Pura Kaitta; *Hindi* Bhoot Katha; *Mising* Kop-kadera; *Nepali* Thople Pokhari Kachhuwa; *Urdu* Chitra Kachhuwa)

DESCRIPTION Carapace tricarinate, with 3 interrupted keels that have nodose prominences; posterior marginals serrated, especially in juveniles, becoming smooth with growth; plastron notched posteriorly and truncated anteriorly; head large, with short snout; digits entirely webbed; shell black with yellow streaks and wedge-shaped marks; head black with yellow spots; neck grey with cream spots; limbs have black-and-white spots. **DISTRIBUTION** Drainages of the Indus, Ganga and Brahmaputra, from India, Bangladesh, Nepal and Pakistan. **HABITAT AND HABITS** Found in oxbow lakes and other standing bodies of water. Omnivorous, diet including molluscs, crustaceans, fish, insect larvae, grasses, fruits and leaves. Clutches comprise 13–30 eggs, 41–51 x 24–28mm. Incubation period 23–76 days. Hatchlings 35–38mm.

Crowned River Turtle ▪ *Hardella thurjii* 61cm (female); 17.5cm (male)
(*Assamese* Bor Dura; *Bengali* Boro Katha, Kali Katha/Kalo Katha; *Bhind* Kariha, Kariyon; *Nepali* Babune Khole Kachhuwa; *Teor* Kariha, Kariyon; *Urdu* Daryai Kachhuwa)

DESCRIPTION Carapace ellipsoidal, domed, fairly depressed, extensively sutured to plastron; buttresses strongly developed; vertebral keel interrupted; head large; digits entirely webbed; carapace dark brown with grey-black vertebral keel; yellowish-orange band usually present on pleuro-marginal area; plastron yellow, each scute with large, blackish-grey blotch; 4 yellowish-orange stripes on sides of head; limbs brownish-grey, edged with yellow. **DISTRIBUTION** Drainages of Indus, Ganga and Brahmaputra, in northern and north-eastern India, Bangladesh, Nepal and Pakistan. **HABITAT AND HABITS** Found in slow-moving rivers, marshes, estuaries and large standing bodies of water. Clutches comprise 8–16 eggs, 40–56 x 28–36mm. Incubation period 223–273 days. Hatchlings 41–46mm.

POND TURTLES

Arakan Hill Turtle ■ *Heosemys depressa* 24.2cm

DESCRIPTION Carapace elongated, flattened, with vertebral keel; weak pair of lateral keels; front and back marginals serrated; head broad with blunt snout; enlarged scales

on forelimbs; fingers and toes weakly webbed; carapace pale brown with dark streaks; plastron yellow with darker streaks and radiations. **DISTRIBUTION** Chittagong Hill Tracts, south-eastern Bangladesh; also adjacent western Myanmar. **HABITAT AND HABITS** Found in hilly rainforests, occupying both streams and leaf litter. Diet comprises fallen fruits and vegetables in the wild; insects, worms and fish accepted in captivity. Clutches comprise 2–6 eggs. Incubation period 120–130 days.

Tricarinate Hill Turtle ■ *Melanochelys tricarinata* 17.5cm
(*Bengali* Shila Kochchop; *Nepali* Padani Kachhuwa, Thotari; *Tindharke* Pahadi Kachhuwa)

DESCRIPTION Carapace elongated, tricarinate, keels low; shell arched with steep sides; snout short, truncate; fingers half webbed; toes almost free; outer surface of forelimbs has

enlarged scales; carapace dark olive, grey-black or reddish-brown, with pale yellow keels; plastron yellow or orange; head and limbs grey-black; yellow or red stripe may extend from nostrils to neck. **DISTRIBUTION** Himalayan foothills of northern India, to north-eastern India, Bangladesh, Bhutan and Nepal. **HABITAT AND HABITS** Terrestrial; associated with grassland, as well as adjacent hills. Crepuscular and omnivorous, feeding on fruits, earthworms, crabs, millipedes, beetles and termites. Clutches comprise 1–3 eggs, 38–47 x 23–25mm. Incubation period 60–72 days. Hatchlings 33–40mm.

▪ Pond Turtles ▪

Indian Black Turtle ▪ *Melanochelys trijuga* 38.3cm
(*Gujarati* Kala Rang No Kachabo; *Hindi* Kala Kachua, Talao Kachua; *Kannada* Kare Aame, Murkate Aame, Neer Aame; *Khasia* D'kar; *Marathi* Kasav; *Nepali* Kalo Pani Kachhuwa, Thotori; *Sinhalese* Gal Ibba, Goo Ibba, Mada Kakkotta, Thumba Ibba, Valan Gibba; *Tamil* Kal Amai, Karuppu Amai, Kullathamai, Neer Amai, Pee Amai, Tanni Amai; *Telugu* Nuiye Tabelu)

DESCRIPTION Carapace elongated, tricarinate, fairly high in adults, depressed in juveniles; head moderate with short snout; upper jaw notched; toes fully webbed; carapace typically brown; plastron usually dark with pale yellow border that may be lost in old individuals; head colour variable, and forms basis of subspecific differentiation.

DISTRIBUTION Northern, peninsular and north-eastern India, Bangladesh, Bhutan, Maldives (probably introduced) and Nepal. Also Myanmar and Thailand. **HABITAT AND HABITS** Found in still waters with aquatic vegetation, and may also occur in small rivers. Omnivorous, eating prawns, grass, water hyacinth and fallen fruits, and scavenges far from water on dead animals. Clutches comprise 3–7 eggs, 44–49 x 26–28mm. Incubation period 76–149 days. Hatchlings 38–44mm.

Brown Roofed Turtle ▪ *Pangshura smithii* 22cm (female); 13cm (male)
(*Bengali* Boro Kori Katha, Vaittal katha; *Hindi* Chapant/Chapatua; *Haire* Dhuri Kachhuwa; *Nepali* Kurdu Bhoora Daryai Kachhuwa)

DESCRIPTION Carapace elliptical, arched, with either smooth or slightly serrated posterior rim; medial keel weak, raised at posterior of vertebral scute; plastron long and narrow; second and fifth vertebrals broader than long; first, third and fourth vertebrals longer than broad; fourth vertebral tapering and pointed anteriorly. In subspecies *smithii*, carapace light green (males) or brown (females); plastron and bridge yellow with single large blotch on scute; head yellowish-grey or pinkish-grey, with distinct brick-red patch on temple; neck grey with yellow stripes; limbs grey. Subspecies *pallidipes* has reduced pigmentation on limbs and head, and pale yellow, unpatterned plastron, though there may be small black smudges on some marginals. **DISTRIBUTION** Pakistan, northern to north-eastern India, Bangladesh and Nepal. **HABITAT AND HABITS** Restricted to large and medium-sized rivers and their vegetation-choked backwaters. Consumes both water plants and fish. Clutches comprise 7–9 eggs, 22–25 x 40–42mm. Hatchlings 35.8mm.

POND TURTLES

Assam Roofed Turtle ■ *Pangshura sylhetensis* 20cm
(*Assamese* Phulen Dura; *Bengali* Kath Kathua, Sylhet Kori Kathua)

DESCRIPTION Carapace elevated, spike-like, especially in juveniles, serrated posteriorly even in adults; the only Indian freshwater turtle with 13 marginals; snout slightly projecting; upper jaw weakly hooked; carapace olive-brown with pale brown vertebral keel; plastron has large black blotches; red sinuous mark runs from eyes to middle of back of head, and another such mark runs along mandibles, curving to join tympanum; neck light striped. **DISTRIBUTION** Eastern and north-eastern India and Bangladesh. **HABITAT AND HABITS** Found in streams, small rivers in hills and foothills, and oxbow lakes. Clutches comprise 6–12 eggs, 34–45 x 21mm.

Indian Roofed Turtle ■ *Pangshura tectum* 23cm
(*Assamese* Futuki Salika Dura; *Bengali* Dora Kaitta, Kori katha; *Gujarati* Rangin Kachabo; *Hindi* Chandan Kachua, Pachauria, Tilhara; *Nepali* Darkhechuwa; *Urdu* Ari-pusht Daryai Kachhuwa)

DESCRIPTION Carapace elevated, oval, with distinct keel on third vertebral that is spike-like, especially in juveniles; head small with projecting snout; upper jaw unnotched, serrated; digits entirely webbed; first vertebral as long as wide or longer than wide; fourth vertebral longer than wide, flask shaped; skin of forehead has irregular scales; carapace brownish-olive, with light brown, red or orange stripe along first 3 vertebral scutes; marginals have narrow yellow border; plastron yellow or pink, with 2–4 black markings on plastral scutes; head has orange or red crescentic postocular markings from below eyes to forehead; neck dark grey with thin yellow or cream stripes. **DISTRIBUTION** Northern India, Bangladesh, Nepal and Pakistan. **HABITAT AND HABITS** Found in standing waters with macrophytes. Basks communally, and several turtles are usually seen basking on logs on sides of river. Omnivorous; juveniles more carnivorous than adults, which turn herbivorous. Clutches comprise 5–10 eggs, 37–45 x 21–24mm. Incubation period 70–78 days. Hatchlings 34–35mm.

■ SOFTSHELL TURTLES ■

> **TRIONYCHIDAE (SOFTSHELL TURTLES)**
> The trionychids include some of the largest freshwater turtles, with 32 living species. They are referred to as 'softshell turtles' due to the absence of scutes on their shells, the bones being clad with skin. The scientific name refers to their three-clawed limbs, another unique trait. They also have long necks and nostrils set on a fleshy proboscis.

Narrow-headed Softshell Turtle ■ *Chitra indica* 1.5m

(*Assamese* Baghia Kasso; *Bengali* Chitra, Chhim, Dhush Kachim, Gotajil, Shuwa Kasim, Thal Kasim; *Hindi* Chitra, Seem, Seonthar, Sewteree; *Nepali* Badar/Badhar, Chitra; *Oriya* Balera Kaichha, Chitra Kaichha; *Urdu* Tang-sar Prait)

DESCRIPTION Carapace oval and depressed; head extremely narrow; eyes small; shell of juveniles has numerous small tubercles and vertebral keel that disappears with growth; dorsum dull olive or bluish-grey, with dark wavy reticulation; carapace pattern continues to neck and outer surfaces of forelimbs; 'V'-shaped mark on nape extends to carapace; juveniles sometimes have 4 eye-like markings on carapace, or numerous black elongated spots; plastron cream or pale pink; head olive with dark-bordered yellow streaks.
DISTRIBUTION Peninsular, northern and north-eastern India, Bangladesh, Nepal and Pakistan.
HABITAT AND HABITS Found in sandy sections of rivers, and ambushes prey underwater by burying itself in sand. Diet includes fish and molluscs. Eggs laid between end of August and mid-September, in sandy or sandy-loam soils. Clutches comprise 65–193 eggs, 25–28mm. Incubation period 40–70 days. Hatchlings 39–43mm.

Sri Lankan Flapshell Turtle ■ *Lissemys ceylonensis* 37cm

(*Sinhalese* Alu Ibba, Kiri Ibba; *Tamil* Pal Amai)

DESCRIPTION Carapace oval, with 7 callosities on plastron; skin-clad, hinged anterior lobe of plastron closes completely; paired plastral flaps; carapace unpatterned greyish-olive; plastron cream coloured; forehead dark grey, lacking distinct patterns.
DISTRIBUTION Sri Lanka. **HABITAT AND HABITS** Inhabits rivers, ponds, rice fields and canals in cities. Diet includes invertebrates as well as fish and frogs.

SOFTSHELL TURTLES

Indian Flapshell Turtle ■ *Lissemys punctata* 37cm
(*Assamese* Bagh Dura, Baminy Kasso; *Bengali* Chip Kathua, Chiti Kachim, Til Kachim; *Gujarati* Pani No Kachbo; *Hindi* Abhua, Matia, Sundri; *Kannada* Boli Aame, Laai Aame; *Malayalam* Vellayama; *Oriya* Panka Kaichha; *Tamil* Pal Amai; *Telugu* Neeti Tabelu; *Urdu* Moonji Kachhuwa)

DESCRIPTION Carapace oval, with 7 callosities on plastron; skin-clad, hinged anterior lobe of plastron closes completely; paired plastral flaps; carapace olive-green with dark yellow blotches in northern subspecies *andersoni*, unpatterned in southern *punctata*; plastron cream or pale yellow. **DISTRIBUTION** India, Bangladesh, Nepal and Pakistan; also Myanmar. **HABITAT AND HABITS** Found in salt marshes, rivers, ponds, oxbow lakes, streams, rice fields and canals in cities. Active by day and night, feeding at dusk. Diet includes frogs, tadpoles, fish, crustaceans, snails, earthworms, insects, carrion and water plants; may also scavenge. Clutches comprise 5–14 eggs, 24–30mm. Southern subspecies produces clutches of 2–8 eggs, 25–33mm. Incubation period 9 months. Hatchlings 42mm.

Carapace

Plastron

Indian Softshell Turtle ■ *Nilssonia gangeticus* 94cm
(*Bengali* Ganga Kachim, Kholua, Kocha; *Gujarati* Kachher, Moti Kachab; *Hindi* Khatawa, Patal; *Nepali* Abhinasha, Ghidiya, Kachhuwa, Seto Bahar; *Oriya* Bada Pani Kaichha, Chabeda Kaichha; *Urdu* Prait)

DESCRIPTION Carapace low and oval; snout slightly downturned; upper jaw without ridges; carapace has longitudinal series of warts in juveniles, smooth in adults; carapace greyish-black, grey or green, with darker reticulation; 3–6 eye-like, yellow-bordered markings on dorsum of juveniles; forehead green, with oblique black stripes on top and sides. **DISTRIBUTION** Northern to north-eastern India, Bangladesh, Nepal and Pakistan.

HABITAT AND HABITS Inhabits rivers, ponds, lakes and reservoirs. During droughts, buries itself in mud at bottoms of ponds and lakes. Frequently seen basking on sand banks, especially in winter. Diet includes water plants, invertebrates and vertebrates; also scavenges on carrion. Clutches comprise 13–35 eggs, 30–35mm.

■ SOFTSHELL TURTLES ■

Indian Peacock Softshell Turtle ■ *Nilssonia hurum* 60cm
(*Assamese* Bor Kasso; *Bengali* Bukum, Dhalua Kachim, Dhum Kachim; *Hindi* Morpanchi; *Nepali* Charthari, Katakhiri; *Oriya* Balikuno, Dhum Kaichha; *Urdu* Peeli Prait)

DESCRIPTION Carapace low and oval; head large with snout strongly downturned; distinct longitudinal rows of blunt tubercles on anterior of carapace; several longitudinal rows of tubercles on posterior of carapace; carapace olive with yellow rim; juveniles have 4–6 dark-rimmed, yellow-bordered occelli; plastron light grey; head and limbs olive; forehead has black reticulation and large orange or yellow patches across snout and on sides. DISTRIBUTION Northern and eastern India, Bangladesh and Nepal. HABITAT AND HABITS Found in rivers, ponds and reservoirs, frequently with dense macrophytic vegetation. Diet includes snails, fish and mosquito larvae. Clutches comprise 20–30 eggs. Hatchlings 46mm.

Black Softshell Turtle ■ *Nilssonia nigricans* 91cm
(*Assamese* Laomura; *Bengali* Bostami Kachim, Gazari, Madari)

DESCRIPTION Carapace low and oval; snout slightly downturned; upper jaw without ridges; carapace with longitudinal series of warts in juveniles, smooth in adults; carapace blackish-grey; 4 yellow-bordered, eye-like markings on dorsum of juveniles; forehead olive-grey with pale areas. DISTRIBUTION North-eastern India and Bangladesh. HABITAT AND HABITS Found in rivers, oxbow lakes and tanks, some of which are man-made and associated with religious shrines. Diet includes both animal and plant matter. Clutches comprise 12–38 eggs, 32–36mm. Incubation period 73–108 days. Hatchlings 45–51mm.

SOFTSHELL TURTLES/MARINE TURTLES

Asian Giant Softshell Turtle ■ *Pelochelys cantorii* 1.5m
(*Bengali* Jata Kachim; *Oriya* Adithia Kaichha, Baligad/Baligado)

DESCRIPTION Carapace low and depressed, and elongated in young, oval in adults; juveniles have tuberculate carapace and low vertebral keel; proboscis short and rounded;

carapace olive or brown, spotted or streaked with lighter or darker shades, pale margined. **DISTRIBUTION** India and Bangladesh; also Southeast Asia, including the Philippines, and central and southern China. **HABITAT AND HABITS** Found in coastal regions such as sea beaches, and also inland lakes and rivers. Diet includes fish, shrimps, crabs and molluscs, and sometimes aquatic plants. Known to share nesting beaches with Olive Ridley Sea Turtle (see p. 28); approaches nesting beaches from both river and sea. Clutches comprise 24–70 eggs, 30mm. Incubation period 60 days. Hatchlings 42mm.

CHELONIIDAE (MARINE TURTLES)
The family includes six living marine turtles, apart from the Leatherback Sea Turtle. They have a relatively more fusiform body shape. Adaptations to their aquatic habits are the lack of ability to retract their heads and limbs into their shells, the smooth shell surface and long forelimbs. Typically they come ashore only to nest.

Loggerhead Sea Turtle ■ *Caretta caretta* 1.2m
Sinhalese Kannadi Kasbava, Olu Gedi Kasbava; *Tamil* Kadal Amai, Kili Chondan Amai, Nai Amai, Panguni, Perunthalai Amai, Ettu Panguni Amai)

DESCRIPTION Carapace elongated, with a tapering end; costals 5 pairs, the first contacting the nuchal; 3 or 4 infralabial scutes that lack pores; 13 marginal scutes; carapace reddish-brown, plastron yellowish-brown or yellowish-orange. Often confused with Olive Ridley Sea Turtle (see p. 28), but can be differentiated in having 5 (rather than 6) costals; bridge with 3 (not 4) inframarginals; and carapace reddish-brown (versus

olive-green or greyish-olive). **DISTRIBUTION** Gulf of Mannar, India, Bangladesh and Sri Lanka. **HABITAT AND HABITS** Inhabits warm, subtropical seas, and known from bays, lagoons and estuaries. Diet consists of molluscs and crustaceans, the large jaws being adapted to crush their shells. Clutches comprise 23–178 eggs, 35–55mm, Incubation period 49–80 days. Hatchlings 44mm.

▪ Marine Turtles ▪

Green Turtle ▪ *Chelonia mydas* 1.4m
(*Bengali* Sabuj Samudrik Kachim; *Car Nicobarese* Kap-chyoot, Kap-ke; *Central Nicobarese* Kap-ka; *Gujarati* Duryani Kachbi; *Hindi* Dudh Kachua, Samundrik Kachua; *Lakshadweep* Mirigham; *Malayalam* Kadal Aamah; *Oriya* Samudra Kaichha; *Sinhalese* Gal Kasbava, Mas Kasbava, Vali Kasbava; *South Nicobarese* Kauske; *Tamil* Pal Amai, Perr Amai, Thaen Amai)

DESCRIPTION Carapace heart shaped; scutes of carapace not overlapping; upper jaw lacks hook; forelimbs have single claw; carapace olive or brown, usually with dark radiating pattern; plastron pale yellow; adult males smaller than females, and possess relatively longer tails that project out of carapace rim. **DISTRIBUTION** Coastal India, including the Andaman Islands, Bangladesh, Pakistan and Sri Lanka. **HABITAT AND HABITS** Widely distributed in tropical regions, and common around oceanic islands and along coasts with wide sandy beaches. Juveniles carnivorous, while adults consume only seagrass and seaweed. Clutches comprise 98–172 eggs, 41–42mm. Incubation period 60 days.

Hawksbill Sea Turtle ▪ *Eretmochelys imbricata* 1m
(*Car Nicobarese* Kap-sah; *Central Nicobarese* Kap-kael; *Hindi* Kanga Kachua; *Oriya* Baja Thantia Kaichha, Chhanchana Tantia Kaichha; *Sinhalese* Pothu Kesbawa; *South Nicobarese* Kaengay; *Tamil* Alunk Amai, Nanja Amai, Ot Amai, Seep Amai)

DESCRIPTION Carapace heart shaped; 4 pairs of overlapping costal scutes; 2 pairs of prefrontal scales; upper jaw relatively narrow and elongate; upper jaw forwards projecting, to form bird-like beak; carapace olive-brown; juveniles have darker blotches than adults. **DISTRIBUTION** India, including the Andaman Islands, Bangladesh, Pakistan and Sri Lanka. **HABITAT AND HABITS** Associated with coral reefs, bays, estuaries and lagoons. Main diet consists of sponges, though algae, corals and shellfish are also eaten. Nesting varies with locality: August and January in the Andamans; June and October on Indian mainland. Clutches comprise 96–177 eggs, 30–35mm. Incubation period 57–65 days.

MARINE TURTLES/AGAMID LIZARDS

Olive Ridley Sea Turtle ■ *Lepidochelys olivacea* 80cm
(*Bengali* Faki Kochchop, Gola Kochchop, Samudrik Katha; *Central Nicobarese* Kap-ngal; *Gujarati* Daryani Kachbi; *Hindi* Gadha Kachua; *Kannada* Kardelu Aame; *Lakshadweep* Malaya Mirigham; *Oriya* Luni Kaichha, Soil Kaichha; *South Nicobarese* Karasara; *Tamil* Kadal Amai, Paingili Amai, Pul Amai, Sith Amai; *Telugu* Samudram Tabelu)

DESCRIPTION Carapace broad, heart shaped, posterior marginals serrated, with juxtaposed costal scutes; 5–9 pairs of costals; bridge with 4 inframarginals, each with pore; adult shell smooth; hatchling shell tricarinate, lateral and with vertebral keels that disappear with growth; upper jaw hooked, but lacks ridge; carapace olive-green or greyish-olive; plastron greenish-yellow; juveniles grey-black dorsally; cream coloured ventrally. **DISTRIBUTION** India, including the Andaman Islands, Bangladesh, Pakistan and Sri Lanka. **HABITAT AND HABITS** Nesting takes place in large aggregations (referred to as 'arribadas'), such as those seen on Gahirmatha and Rushikulya beaches, Odisha, on the east coast of India, where several hundred thousand turtles congregate to nest. Clutches comprise 50–160 eggs, 34–43mm. Incubation period 45–60 days. Hatchlings 38–50mm.

AGAMIDAE (AGAMID LIZARDS OR DRAGONS)
These sit-and-wait predators are part of a large lizard family (463 described species) from southern Europe, Africa, Asia and Australasia, which is related to the New World iguanas. They have well-developed limbs, non-autotomous tails, some capacity to change colours for social signalling or regulation of body temperature, and acrodont teeth (set on outer rim of mouth). They are diurnal and generally terrestrial or arboreal (the Laungwala Toad-headed Lizard being the only specialized burrower). Their diets range from arthropods such as insects and spiders, to birds and mammals. A few large-growing species include flowers, seeds and leaves in their diets. Most species are oviparous.

◾ Agamid Lizards ◾

Laungwala Toad-headed Lizard ◾ *Bufoniceps laungwalansis* 67mm
(*Rajasthani* Girgit, Chipkali)

DESCRIPTION Body short and depressed; tiny external ear opening and deeply set tympanum; snout short, with nostrils set close together high on snout; dorsal scales small, uniform; tail short; no dorsal crest, gular sac, or femoral and preanal pores; digits bear fringes of flat, pointed scales; dorsum grey with red, orange, black and white spots; dark vertebral stripe sometimes present; distinct blue patch between shoulder and neck; belly white or cream coloured. **DISTRIBUTION** Western India and eastern Pakistan. **HABITAT AND HABITS** Found in sandy deserts at 139–290m. Diurnal and active on subsurface of loose sand of shifting sand dunes, running for short distances and rapidly burrowing 2–3cm in sand, using lateral movements. Diet comprises ants, beetles, grasshoppers, flies and lizards. Reproductive habits unknown.

Lateral view

Dorsal view

Green Forest Lizard
◾ *Calotes calotes* 130mm
(*Sinhalese* Kola Katussa, Pala Katussa; *Tamil* Pachchai Onnan, Pachchonthi)

DESCRIPTION Body robust; head large; cheek swollen in adult males; crest on head and body distinct; oblique fold in front of shoulder; throat sac not well developed; dorsal scales smooth or weakly keeled, pointing backwards and upwards; tail long and rounded; dorsum bright green, with 4–5 bluish-white or green cross-bars; belly pale green. **DISTRIBUTION** Southern peninsular India and Sri Lanka. **HABITAT AND HABITS** Found in moist deciduous and evergreen forests in mid-hills and plains. Arboreal; associated with shrubs and tree trunks. Diet consists of insects. Clutches comprise 6–12 eggs, 12–12.5 x 18–18.5mm. Incubation period 79–84 days.

▪ AGAMID LIZARDS ▪

Painted-lipped Lizard ▪ *Calotes ceylonensis* 70mm
(*Sinhalese* Thola-visituru Katussa)

DESCRIPTION Body slender, compressed; head one and a half times width; tail long, slender, over twice head-body length; cheeks swollen; 2 separated spines above tympanum; nuchal crest of low spines; midbody scale rows 54–60; dorsum brown with indistinct brown bands; head and anterior part of body blackish-brown; bright red or reddish-orange stripe on upper lip extends to back of head; red dorsonuchal crest; throat of adult males black;

tail brown with dark bands; belly pale brown with darker bands. **DISTRIBUTION** Sri Lanka. **HABITAT AND HABITS** Found in semi-evergreen forests, plantations and home gardens in dry and intermediate zones, at 400m. Arboreal; active on tree trunks. Diet includes insects and other arthropods. Clutches comprise 5–10 eggs, 14.5 x 16.5mm.

Emma Gray's Forest Lizard ▪ *Calotes emma* 115mm

DESCRIPTION Body robust; head short; dorsal scales point backwards and upwards; cheek swollen in adult males; large spine above eyes and 2 above tympanum; fold in front

of shoulder; nuchal and dorsal crests continuous; dorsum olive-brown, with dark brown dorsal bars or transverse spots; dark radiating lines from eyes; red gular sac in breeding males; pale dorsolateral stripe; belly greyish-cream. **DISTRIBUTION** North-eastern India and Bangladesh; also southern China and Southeast Asia. **HABITAT AND HABITS** Found in forested mid-hills. Diet includes insects. Clutches comprise 4–12 eggs, 11 x 17mm.

◾ AGAMID LIZARDS ◾

Jerdon's Forest Lizard ◾ *Calotes jerdoni* 100mm
(*Bengali* Sabuj Raktachosha; *Angami Naga* Sokru)

DESCRIPTION Body robust; head large; dorsal crest present; nuchal crest weak; distinct fold in front of shoulder; dorsal scales larger than ventrals; 2 parallel rows of compressed scales above tympanum; dorsum bright green, with paired, black-edged brown bands, and yellow, orange or brown blotches; black and brown morphs also known; tail dark banded or spotted. DISTRIBUTION North-eastern India and Bhutan; also southern China and Southeast Asia. HABITAT AND HABITS Inhabits forested low hills. Diet includes insects. Clutches comprise 12 eggs. Hatchlings 70mm.

Whistling Lizard ◾ *Calotes liolepis* 85mm
(*Sinhalese* Kalae Katussa, Sivuruhandalana Katussa)

DESCRIPTION Body stout; head elongated; tail long, slender, nearly 2.5 times head-body length; series of spines on nape make up dorsonuchal crest in males; enlarged scales on dorsum of body; midbody scale rows 33–39; ventral scales as large as those on flanks; forehead pale brown with pale interorbital bands; dorsum pale grey with dark grey bands, numbering 4 on body; limbs and tail similarly dark banded. DISTRIBUTION Central hills of Sri Lanka. HABITAT AND HABITS Found in submontane forests below 1,000m. Diet comprises insects and ants. Reproductive habits unknown.

◾ AGAMID LIZARDS ◾

Maria's Lizard ◾ *Calotes maria* 120mm

DESCRIPTION Head large; body stout, compressed; scales on body point backwards and upwards; 2 parallel rows of compressed scales above tympanum; gular pouch absent;

nuchal and dorsal crests developed; dorsum green, with red streaks on flanks and red spots on limbs and tail; head of breeding males red; belly greenish-white. **DISTRIBUTION** North-eastern India and Bhutan; also Myanmar. **HABITAT AND HABITS** Found in low forested hills. Diurnal and arboreal. Diet comprises arthropods. Reproductive habits unknown.

Moustached Forest Lizard ◾ *Calotes mystaceus* 140mm
(*Manipuri* Numityoungbi Chum)

DESCRIPTION Body robust; head large; cheeks swollen; 2–3 spines behind eye; fold in front of shoulder; dorsum greyish-brown, turning bright blue to turquoise, with 3–5

large dark spots on sides, during breeding season; lips white; belly greyish-cream. **DISTRIBUTION** North-eastern India; also Southeast Asia. **HABITAT AND HABITS** Found in evergreen forests, entering parks and gardens. Diet includes insects. Clutches comprise 7 eggs, 10–11 x 15–18mm. Incubation period 60 days. Hatchlings 26mm.

Agamid Lizards

Black-lipped Lizard — *Calotes nigrilabris* 105mm
(*Sinhalese* Kaludekupul Katussa, Kalu Kopul Katussa)

DESCRIPTION Body robust; head one and a half times width; row of spines above and at back of tympanum; adult males have swollen cheeks; gular sacs not developed; midbody scale rows 42–50; ventrals larger than dorsals; dorsum green, unpatterned or with black-edged, cream-coloured transverse bars or eye-like spots; head has black markings; belly pale green. **DISTRIBUTION** Sri Lanka. **HABITAT AND HABITS** Found in submontane forests at 1,000m and above. Associated with tree trunks, hedges and shrubs. Diet consists of insects and worms. Clutches comprise 2 eggs, 23 x 13mm. Hatchlings 30mm.

Garden Lizard — *Calotes versicolor* 140mm
(*Assamese* Tejpia; *Bengali* Girgiti, Raktachosha; *Dhivehi* Bondu; *Gujarati* Dol Kanchido, Sarado, Saradi; *Kannada* Hente Goda; *Hindi* Girgit; *Malayalam* Ohnthoo; *Manipuri* Numityoungbi Chum; *Marathi* Sarda/Seda; *Oriya* Endua; *Rajasthani* Kangetia/Kirkanthia; *Sindhi* Shyee; *Sinhalese* Gara Katussa; *Tamil* Onnan, Wona; *Telugu* Thonda, Thota Balli; *Urdu* Girgit/Girgitan, Kafir Girgit)

DESCRIPTION Body stout; head rather large; scales on body point backwards and upwards; no fold or pit in front of shoulders; 2 separated spines above tympanum; colouration variable and also changeable, the head becoming bright red, and black patch on throat appearing in displaying males, fading to dull grey at other times; females may become yellow, changing to dull greyish-olive after mating. **DISTRIBUTION** India, Bangladesh, Bhutan, the Maldives, Nepal, Pakistan and Sri Lanka; also Iran, Afghanistan and Southeast Asia. **HABITAT AND HABITS** Found in forests as well as shrubs and hedges around human landscapes. Diet consists of insects and other invertebrates; unripe seeds also consumed. Clutches comprise 6–23 eggs, 10–11 x 4–5mm. Incubation period 42–67 days.

Male in display (NE India)

Male in display (S India)

AGAMID LIZARDS

Rough-horned Lizard ■ *Ceratophora aspera* 35mm
(*Sinhalese* Punchi Ang Katussa, Ralu Ankatussa)

DESCRIPTION Body slender; 2 enlarged, conical ridged scales at back of head; 'X'-shaped dorsal ridge at back of head; tympanum hidden; weak dorsonuchal crest confined to neck region; gular fold absent; body slightly compressed; lamellae under fourth toe 11–14; dorsum of males dark brown or brick-red; females similar or lighter; some individuals have 4 diamond-shaped marks, and black spots or longitudinal lines on dorsum. **DISTRIBUTION** Sri Lanka. **HABITAT AND HABITS** Found in moist lowlands and submontane dipterocarp forests, below 900m. Diet unknown. Clutches comprise 2 eggs.

Rhinoceros-horned Lizard
■ *Ceratophora stoddartii* 85.6mm
(*Sinhalese* Kagamuva Angkatussa, Rhino Angkatussa)

DESCRIPTION Body slender; head oval, longer than wide; rostral appendage long, horn-like, and two-thirds length of snout in males, reduced or absent in females; dorsum brownish-green or yellowish-brown; tail has 10–16 dark brown bands; belly light brownish-grey. **DISTRIBUTION** Sri Lanka. **HABITAT AND HABITS** Found in montane forests. Associated with trees 1–2m above ground. Diet consists of arthropods. Clutches comprise 2–5 eggs, 7.6–8 x 13.5–14.5mm. Incubation period 81–90 days.

= Agamid Lizards =

Leaf-nosed Lizard ■ *Ceratophora tennentii* 88.5mm
(*Sinhalese* Pathra Angkatussa, Pethi Angkatussa)

DESCRIPTION Body slender; head oval, longer than wide; rostral appendage fleshy, laterally compressed, with blunt conical scale at tip; dorsum reddish-brown to olive-green; flank scales more green; ocular region and sides of neck have dark markings; tail has 10 dark brown bands; belly cream coloured. **DISTRIBUTION** Knuckles Massif of Sri Lanka. **HABITAT AND HABITS** Found in submontane forests at 760–1,220m. Arboreal, associated with tree trunks. Diet includes insects and their larvae. Clutches comprise 3–4 eggs.

Pygmy Lizard ■ *Cophotis ceylanica* 60mm
(*Sinhalese* Kandukara Katussa, Kuru Bodiliya)

DESCRIPTION Body stout, compressed; head narrow; long dorsonuchal crest developed; temporal scales with 3–5 large, conical scales; tympanum absent; tail short and prehensile; dorsal scales enlarged; gular sac laterally compressed; preanal and femoral pores absent; dorsum olive-green, with darker markings forming 3 bands on body and more on tail; light spot on nape; broad stripe along anterior of body and one in front of eyes; limbs dark banded. **DISTRIBUTION** Sri Lanka. **HABITAT AND HABITS** Found in montane regions at 1,300–1,900m elevation. Diet unknown. Clutches comprise 4–5 live young, 47–50mm.

◾ Agamid Lizards ◾

Bay Islands Forest Lizard ◾ *Coryphophylax subcristatus* 100mm
(*Karen Po-tenu*)

DESCRIPTION Body slender; dorsal crest present; dorsal scales small, intermixed with larger scales; preanal and femoral pores absent; cheeks swollen in adult males; ventral scales strongly keeled; dorsum brownish-olive, unpatterned, spotted or striped with dark brown; juveniles more green than adults, with dark cross-bars; belly light brown. **DISTRIBUTION** Andaman and Nicobar Islands, India. **HABITAT AND HABITS** Found in lowland rainforests and plains, including edges of mangrove forests. Diet consists of insects. Clutches comprise a single large egg.

Norville's Flying Lizard
◾ *Draco norvilii* 96mm

DESCRIPTION Body slender; nostrils directed upwards; tympanum covered with small scales; 5 ribs supporting wing membrane; males have tail crest; dorsum greenish-olive to pale green; wing membrane pale green towards body, darker on edges, with wide, dark grey bands; peripheral third of wing membrane orangish-red; belly yellow; gular pouch orangish-yellow; base grey with light spots at border with lateral pouches. **DISTRIBUTION** North-eastern India; also adjacent Myanmar. **HABITAT AND HABITS** Inhabits evergreen and moist deciduous forests, and also plantations. Diet and reproductive habits unknown.

AGAMID LIZARDS

Anderson's Mountain Lizard ■ *Japalura andersoniana* 75mm

DESCRIPTION Body slender, compressed; tail long; tympanum concealed; enlarged scales on dorsum parallel to median row; nuchal crest distinct in adult males; throat-fold present; dorsum dark brown with dark, inverted 'V'-shaped pattern; dark ocular region; tan streak from eye to edge of mouth; throat of males variable, and may be orange confined within green spot, or blue-and-white stripes, with yellow with orange or green spot; belly cream coloured. **DISTRIBUTION** North-eastern India; also southern China. **HABITAT AND HABITS** Found in montane forests in hilly country, at 850–1,980m. Diet comprises insects. Reproductive habits unknown.

Kumaon Mountain Lizard ■ *Japalura kumaonensis* 60mm
(*Urdu* Kumaon Kirail)

DESCRIPTION Body slender, somewhat compressed; tail long; postorbital spine absent; tympanum exposed; preanal and femoral pores absent; enlarged scales on dorsum parallel to median row; throat-fold absent; dorsum greyish-brown with dark, inverted 'V'-shaped marks; sides with dark reticulation; forehead has dark cross-bars; belly cream coloured, sometimes with darker markings. **DISTRIBUTION** Northern India and Nepal. **HABITAT AND HABITS** Found in montane forests in hilly country. Diet comprises insects. Reproductive habits unknown.

◾ Agamid Lizards ◾

Flat-backed Mountain Lizard ◾ *Japalura planidorsata* 50mm

DESCRIPTION Body slender, not compressed; dorsal and nuchal crests comprise spinose ridges; tail rounded; tympanum concealed; short axillary fold present; dorsum pale to mid-brown, paler on vertebral region; lips cream or yellow; belly pale brown or cream. **DISTRIBUTION** North-eastern India; also Myanmar. **HABITAT AND HABITS** Found in forested hills and plains. Diet comprises insects. Reproductive habits unknown.

Kashmiri Rock Agama ◾ *Laudakia tuberculata* 140mm
(*Nepali* Cheparo, Pahari Chalwar; *Urdu* Neela Kirla)

DESCRIPTION Body robust, flattened; throat scales keeled; patch of enlarged scales on mid-flanks; tail has distinct segmentation, each segment composed of double whorls of scales; dorsum dark olive-brown, dark-spotted in juveniles, with spots broken up to form speckled pattern of dark brown and yellow; chest, shoulder and flanks have orange or yellow spots; belly brown or cream coloured. **DISTRIBUTION** Northern India, Nepal and Pakistan; also Afghanistan. **HABITAT AND HABITS** Found in rock crevices, in colonies or singly. Breeding males develop bright blue colouration, when they bob their heads. Diet consists of insects, spiders, millipedes, centipedes, butterflies and flower petals. Clutches comprise 6–13 eggs, 20–22 x 11–12mm. Incubation period 31 days.

▪ Agamid Lizards ▪

Hump-nosed Lizard ▪ *Lyriocephalus scutatus* 185mm
(*Sinhalese* Karamal Bodiliya, Kandhu Bodiliya)

DESCRIPTION Body laterally compressed; distinct bony arch on head of adults, with pair of small spines; dorsonuchal crest developed; forehead scales keeled; tympanum absent; 'V'-shaped gular fold; large, keeled gular scales; tail short, compressed, with blunt tip; dorsum light green, throat yellow, belly cream coloured. **DISTRIBUTION** Sri Lanka. **HABITAT AND HABITS** Found in forests in wet lowlands and mid-hills, at 25–1,650m. Arboreal and semi-terrestrial; occurs on low trees as well as on the ground. Diet consists of earthworms, termites, butterflies and moths, and young shoots and buds. Clutches comprise 1–11 eggs, 12–13 x 20–22mm. Incubation period 35–36 days.

Sri Lankan Kangaroo Lizard ▪ *Otocryptis wiegmanni* 69mm
(*Sinhalese* Kala Katussa, Pinun Katussa, Thalli Katussa, Yak Katussa)

DESCRIPTION Body compressed; snout blunt; eyes large; weak dorsal crest on neck only; males have large gular sac but no gular fold; shallow pit in front of shoulder; tympanum hidden; limbs long and slender; tail rounded, long, nearly 2.5 times length of head and body; femoral pores absent; dorsum brown with lighter and darker markings. **DISTRIBUTION** Sri Lanka. **HABITAT AND HABITS** Found in lowlands and mid-hills, to 1,200m. Associated with leaf litter, particularly near forest streams. Diet includes insects and their larvae, and vegetation such as tender shoots. Clutches comprise 3–5 eggs, 7–7.5 x 10–17mm. Incubation period 57–70 days.

▪ AGAMID LIZARDS ▪

Caucasian Agama ▪ *Paralaudakia caucasia* 153mm
(*Urdu* Kohkaf Ka Kirla)

DESCRIPTION Body robust; throat scales smooth; patch of enlarged scales on midflanks; tail has distinct segmentation, with each segment composed of double whorls of scales; dorsum olive, dark brown or yellow-ochre; head and tail lighter; dorsum has numerous dark-edged orange occelli that are indistinct posteriorly; throat of males spotted

with bright yellow; belly unpatterned dark grey. **DISTRIBUTION** Northern India and Pakistan; range extends west to Turkey, Central Asia and Iran. **HABITAT AND HABITS** Found in rocky outcrops, including cliffs and boulders along river banks. Diurnal, basking on hot rocks during day. Diet consists of lizards and insects. Clutches comprise about 12 eggs.

South Indian Rock Agama ▪ *Psammophilus dorsalis* 135mm
(*Tamil* Thendel)

DESCRIPTION Body robust, depressed; dorsal crest or gular sac absent; head large; limbs small; scales uniform, keeled; deep fold in front of shoulder; similar to **Blanford's Rock**

Agama *P. blanfordanus*, except has 115–150 scales around middle of body and is larger; dorsum of adult males brown, with dark brown or black stripe along flanks; belly yellow; juveniles and adult females olive-brown, with dark brown spots and speckles and white areas on sides of neck; upper body of adult males bright red or orange in breeding season. **DISTRIBUTION** Peninsular India. **HABITAT AND HABITS** Found in rocky biotopes in scrub country, as well as deciduous forests. Diet includes insects. Clutches comprise 7–8 eggs, 6 x 12mm.

■ Agamid Lizards ■

Green Fan-throated Lizard ■ *Ptyctolaemus gularis* 80mm

DESCRIPTION Body slender, compressed; head long and slender; dorsal scales keeled; dorsal crest absent; 3 longitudinal folds on each side of throat curve to meet on back; gular sac; dorsum olive-brown; fold on back deep blue; 5 broad transverse bands on body; green dorsolateral band on front flanks; sides have network of dark brown enclosing rounded areas of green. **DISTRIBUTION** North-eastern India, Bhutan and Bangladesh; also southern China. **HABITAT AND HABITS** Found in submontane and lowland forests. Arboreal and diurnal. Diet consists of insects, spiders and soil arthropods. Clutches comprise 14–15 eggs, 7–12.5mm.

Hardwicke's Spiny-tailed Lizard ■ *Saara hardwickii* 350mm
(*Gujarati* Sandho, Sandno; *Hindi* Mehta, Sanda; *Punjabi* Salma, Sana, Sonder; *Sindhi* Sonder; *Urdu* Maidani Sanda)

DESCRIPTION Body stout and depressed; head large, tympanum large; tail thick at base, short, depressed, covered with large, squarish spinose scales, largest on sides; front teeth large; preanofemoral pores 12–18; dorsum yellowish-brown, with indistinct brown spots and reticulation; radiating dark streaks from eyes; belly cream coloured. **DISTRIBUTION** Western India and eastern Pakistan. **HABITAT AND HABITS** Found in desert edges, semi-deserts and scrub forests, especially on hard ground with compact loess. Diurnal and terrestrial as well as fossorial, burrowing up to 3m deep. Diet consists of grasses, and insects such as locusts and beetles. Clutches comprise up to 15 eggs, 20–30mm.

Agamid Lizards

Horsfield's Spiny Lizard ■ *Salea horsfieldii* 250mm

DESCRIPTION Body robust, compressed; head large; dorsal crest distinct, discontinuous with nuchal crest; no fold in front of shoulder; dorsal scales equal; dorsum green or greyish-

cream, with dark grey or dark brown cross-bars or irregular blotches on dorsum and flanks; dark band, edged with white, runs from eye to shoulder; belly cream coloured, spotted with brown. DISTRIBUTION Western Ghats of south-western India. HABITAT AND HABITS Found in submontane and lowland forests. Arboreal and diurnal. Diet includes insects. Clutches comprise 3–4 eggs.

Sri Lankan Fan-throated Lizard ■ *Sitana bahiri* 60mm
(*Sinhala* Pullibin Katussa, Vali Katussa; *Tamil* Visiri Wona)

DESCRIPTION Body slender; snout rather acute; tympanum present; hindlimbs elongated, with only 4 toes; scales keeled; femoral pores absent; tail long and slender; dewlap large, projecting, in males; gular fold absent; dorsum brown, with dark brown to greyish-brown, black-edged, diamond-shaped marks; tan-coloured stripe from tympanum to above axilla; mouth lining dark blue; dewlap blue on outer face, the rest cream coloured with large orange scales; belly cream coloured. DISTRIBUTION Northern and eastern Sri Lanka. HABITAT AND HABITS Found in coastal scrub forests. Diurnal and terrestrial. Diet includes arthropods. Reproductive habits unknown.

Agamid Lizards

Eastern Fan-throated Lizard
Sitana ponticeriana 80mm
(*Tamil* Sit Wona, Thadi Onnan, Visiri Onnan)

DESCRIPTION Body slender; snout rather acute; tympanum present; hindlimbs have 4 toes; scales keeled; femoral pores absent; tail long and slender; dewlap large, projecting, in males; gular fold absent; enlarged scales of dewlap reach mid-belly; dorsum brown, with dark brown, black-edged, diamond-shaped marks; mouth lining dark blue; dewlap blue on tip, dark blue in middle and red at base; belly cream coloured. **DISTRIBUTION** Peninsular India. **HABITAT AND HABITS** Found in scrub forests and on sea beaches. Diurnal and terrestrial. Diet includes termites, beetles and bugs. Clutches comprise 6–8 eggs, 6 x 10mm.

Brilliant Ground Agama
Trapelus agilis 115mm
(*Sindhi* Karrun; *Urdu* Maidani Korrh-kirla)

DESCRIPTION Body robust; head enlarged; dorsal scales subequal, with no enlarged scales; middorsal scale rows 60–76; dorsum grey, brown or sandy-yellow, with dark brown or red cross-bars, comprising vertebral and 1–2 dorsolateral light oval spots; throat of breeding males blue, their tails bright yellow with dark brown bands; flanks and belly marked with purple; belly of non-breeding males and females cream, streaked with brown. **DISTRIBUTION** Western India and Pakistan; range extends west to Russia, western China, Iran and Iraq. **HABITAT AND HABITS** Found in flat, dry alluvial or stony and sometimes hard soil, with prickly bush and scrub. Diurnal and terrestrial. Diet consists of grasshoppers. Clutches comprise 2 eggs.

AGAMID LIZARDS/CHAMELEONS

Occelated Ground Agama ■ *Trapelus megalonyx* 67mm
(*Urdu* Patta Korrh-kirla)

DESCRIPTION Body stout; dorsals heterogenous, with larger and smaller scales arranged in groups and both smooth and keeled; ventrals weakly keeled, as large as dorsals; dorsum bronze in males; vertebral series of 6 dark-edged brown cross-bars; throat blue; belly cream with light mottling; dorsum of females brown to pale grey, with vertebral rows of cross-bands; throat and belly cream coloured. **DISTRIBUTION** Pakistan; also Afghanistan and possibly Iran. **HABITAT AND HABITS** Inhabits deserts of alluvial soil, with thin vegetation of grass and shrubs. Associated with roots of vegetation. Diet includes plants and arthropods. Clutches comprise 4–6 eggs.

CHAMAELEONIDAE (CHAMELEONS)
Chameleons belong to a family of 203 described species. Many have the ability to change colours – it is famously believed that they use this for camouflage, but it is in fact more commonly used in social signalling or for thermoregulation. Most species have zygodactylous feet. They have independently mobile eyes, long, rapidly extrudable tongues, prehensile tails and, typically, crests or horns on the cranium.

Indian Chameleon ■ *Chamaeleo zeylanicus* 183mm
(*Bengali* Bahurupi; *Gujarati* Dolan Kauchido, *Kannada* Gossambe, Hasuronthi; *Malayalam* Pachai Ohnthu; *Marathi* Girgitan, Pek-sadha; *Oriya* Endua Shapo; *Sinhalese* Bodilima, Bodiliya; *Tamil* Pachchai Wona; *Telugu* Oosarawalli; *Urdu* Taj-sar Girgit)

DESCRIPTION Body laterally compressed; head has helmet-like, bony projection; orbit of eye large; eyeball covered with skin, leaving tiny aperture; low, serrated dorsal crest extending to prehensile tail; fingers and toes opposable; dorsum green to yellow, with spots or bands. **DISTRIBUTION** Western and Peninsular India, Sri Lanka and Pakistan. **HABITAT AND HABITS** Found in dry forests and scrubland. Arboreal, inhabiting shrubs and trees. Diet includes insects. Clutches comprise 10–31 eggs, 16–19 x 10–12.5mm.

▪ TRUE GECKOS ▪

GEKKONIDAE (TRUE GECKOS)
True geckos constitute a large family of 1,097 described species of living lizard that are distributed in warm regions of the world. Most species are nocturnal and insectivorous, lack movable eyelids and soft scales, are capable of loosing the tail as a form of defence, and can then regenerate the lost tail. Several have specialized toe-pads that permit them to scale smooth and vertical surfaces. A majority are oviparous, a few are ovoviviparous and some are even parthenogenetic (the females are capable of producing eggs without mating). Some are unique among lizards in having the ability to vocalize, which is used in social interactions.

Indian Golden Gecko ▪ *Calodactylodes aureus* 89mm

DESCRIPTION Body robust; head wider than body; body covered with small, flat scales, with scattered rounded tubercles; undersurfaces of fingers and toes have plate-like expanded scansors; adult males have 2 preanal pores; femoral pores 1–6; dorsum of adult males bright yellow, especially on throat; females and juveniles olive-yellow, reddish-brown or blackish-brown. **DISTRIBUTION** Peninsular India. **HABITAT AND HABITS** Found in rocky landscapes and scrubland. Diet consists of grasshoppers, beetles, butterflies and their larvae, spiders, ants and lizard eggs. Clutches comprise 2 eggs, 12.3–12.9 x 14.4–14.9mm, laid communally, glued to rock surfaces. Hatchlings 27.9–29.1mm.

■ True Geckos ■

Sri Lankan Golden Gecko ■ *Calodactyloides illingworthorum* 95mm
(*Sinhalese* Maha Gal Huna, Gal Pahuro)

DESCRIPTION Body robust; head wider than body; pupil vertical; 2 pairs of enlarged, nearly rectangular lamellae under each finger or toe; tail has 27 segments; dorsum yellow-ochre, with dark brown spots; throat bright yellow or orange; chest and belly pale grey or yellow. **DISTRIBUTION** South-eastern Sri Lanka. **HABITAT AND HABITS** Found in rocky landscapes, including granitic caves within savannah and monsoon forests. Diet consists of dipterans, coleopterans, their larvae, glow-worms, and other arthropods. Clutches comprise 2 eggs, 14.9 x 8.2mm, glued to rock surfaces. Hatchlings 27mm.

Assamese Day Gecko ■ *Cnemaspis assamensis* 28mm

DESCRIPTION Body slender; head distinct from neck; eyes large, with rounded pupils; ventral scales increase in size from chin to throat; spines on sides of body absent; tubercles on dorsum of body absent; smooth scales on belly; preanal and femoral pores absent; 2–5 enlarged scansors on toes; tail segmented, with flattened scales forming whorls; dorsum light brown with pale vertebral stripe; chevron-like pale pattern on back, dark-edged posteriorly; pale transverse bar at back of head; black nuchal spot; belly cream coloured. **DISTRIBUTION** North-eastern India and Bhutan. **HABITAT AND HABITS** Found in moist-deciduous and subtropical evergreen forests at low elevations. Associated with low tree trunks and rocks. Diet presumably insects. Clutches comprise 2 eggs.

▪ True Geckos ▪

Indian Day Gecko ▪ *Cnemaspis indica* 38mm

DESCRIPTION Body slender; head distinct from neck; snout elongate; eyes large, with rounded pupils; dorsal scales keeled; spines on sides of body absent; tubercles on dorsum of body absent; smooth scales on belly; 4–5 femoral pores; tail segmented, with flattened scales forming whorls; dorsum light brown, with red or orange spots and mottling; belly brownish-cream; throat dark brown. **DISTRIBUTION** Western Ghats of south-western India. **HABITAT AND HABITS** Found in evergreen forests. Associated with tree trunks and rocks. Diet comprises insects. Eggs 7.7 x 6.9mm.

Kandy Day Gecko ▪ *Cnemaspis kandiana* 40mm

DESCRIPTION Body slender; head distinct from body; snout elongated; spinous tubercles on flanks; ventral scales smooth; males have 2–4 preanal and 3–6 femoral pores; dorsum brown, with lighter and darker variegation arranged transversely; spinous tubercles on flanks cream coloured; belly pale brown. **DISTRIBUTION** Sri Lanka. **HABITAT AND HABITS** Found in submontane forests. Diurnal; active in rocky substrates and low trunks of trees, and occasionally encountered in thatched huts and cow-sheds. Clutches comprise 1–2 eggs, 4–6mm. Hatchlings 13.5–15mm.

▪ TRUE GECKOS ▪

Rough-bellied Day Gecko ▪ *Cnemaspis tropidogaster* 40mm

DESCRIPTION Body slender; head distinct from body; snout elongated; spinous tubercles on flanks; ventral scales keeled (not smooth, as in Kandy Day Gecko, see p. 47); males

have 2–4 preanal and 3–6 femoral pores; dorsum dark brown, with lighter and darker variegation arranged transversely; spinous tubercles on flanks cream coloured; belly pale brown. DISTRIBUTION Western Ghats of south-western India and Sri Lanka. HABITAT AND HABITS Found in forested hills. Diurnal; associated with rocky areas and low trunks of trees, and occasionally found in thatched huts and cow-sheds.

Yercaud Day Gecko ▪ *Cnemaspis yercaudensis* 25mm

DESCRIPTION Body slender; head large, distinct from body; snout longer than eye; eyes large, with rounded pupils; 3 pairs of paravertebral rows of tubercles; no spine-like tubercles

on flanks; preanal pores 2; femoral pores 3; dorsum greyish-brown, with dark mottling; belly unpatterned yellowish-cream. DISTRIBUTION Shevaroy Hills of south-western India. HABITAT AND HABITS Found in middle elevations, and known only from a stone wall at edge of road. Diet and reproductive habits unknown.

▪ True Geckos ▪

Sindh Sand Gecko ▪ *Crossobamon orientalis* 53mm
(*Urdu* Pelee-dum)

DESCRIPTION Body stout; head rather large; dorsal scales small, mixed with rounded tubercles; ventral scales small, smooth; fingers and toes have fringe of small, pointed scales; femoral pores sometimes absent in males, and when present may number 1–4; dorsum brownish-yellow to pale grey, with 3–5 indistinct grey bands; tail yellow with distinct dark rings; belly cream coloured. **DISTRIBUTION** Western India and Pakistan. **HABITAT AND HABITS** Found in sand dunes and areas with fine sand and sparse vegetation. Terrestrial and nocturnal. Diet includes termites and other insects. Clutches comprise 3 eggs.

Nicobar Bent-toed Gecko ▪ *Cyrtodactylus adleri* 69mm

DESCRIPTION Body slender; head long; snout tapering; ear opening small, oval; digits slender, recurved, not dilated, with enlarged lamellae on ventral surface; tail with tubercles, and long, tapering to fine point; 6 preanal pores in males; dorsum and forehead mid-brown, forehead with 7 rounded dark blotches; dark postocular stripe to armpit; 2 dark spots on neck region; belly cream coloured. **DISTRIBUTION** Great Nicobar Island, India. **HABITAT AND HABITS** Found in lowland rainforests. Arboreal. Diet and reproductive habits unknown.

True Geckos

Kollegal Ground Gecko ◾ *Cyrtodactylus collegalensis* 47mm

DESCRIPTION Body stout, cylindrical, covered with small, granular scales; scales on belly overlapping; tail short, tapering, regenerated tail fat; dorsum has 5 pairs of large dark

brown spots, in addition to 3 pairs on head; belly cream coloured. **DISTRIBUTION** Western peninsular India. **HABITAT AND HABITS** Found in deciduous and scrub forests. Crepuscular, hiding under rocks by day and emerging at dusk. Diet includes insects such as termites. Clutches comprise 2 eggs, 8.6–8.7 x 10–11mm. Incubation period 43 days. Hatchlings 33.5mm.

Banded Bent-toed Gecko
◾ *Cyrtodactylus fasciolatus* 82mm

DESCRIPTION Body slender; head long; snout tapering; ear opening small, oval; digits slender, recurved, not dilated, with enlarged lamellae on ventral surface; tail has tubercles; tail long and tapering to fine point; dorsum and forehead mid-brown; forehead has 7 rounded dark blotches; dark postocular stripe to axilla; 2 dark spots on neck region; belly cream coloured. **DISTRIBUTION** Western Himalayas of northern India. **HABITAT AND HABITS** Found in submontane forests, up to 1,500m. Arboreal, and also associated with rocks. Diet and reproductive habits unknown.

▪ True Geckos ▪

Khasi Hills Bent-toed Gecko ▪ *Cyrtodactylus khasiensis* 85mm

DESCRIPTION Body slender; snout relatively long; dorsal surface of body and limbs has small, granular scales, intermixed with larger, keeled tubercles; lateral fold of enlarged scales; males have 12–14 preanal pores; femoral pores absent; dorsum dark greyish-brown, with dark spots; curved dark mark extends across nape to meet eyes; forehead black spotted; belly cream coloured. DISTRIBUTION North-eastern India and Bhutan; also northern Myanmar. HABITAT AND HABITS Found in submontane forests. Nocturnal; associated with rocks and low vegetation. Diet consists of insects. Clutches comprise 2 eggs.

Andaman Bent-toed Gecko ▪ *Cyrtodactylus rubidus* 75mm
(*Karen* Daw-ly Phawo)

DESCRIPTION Body slender; head long; snout tapering; ear opening small, oval; digits have enlarged lamellae on ventral surface; tail has tubercles; tail long and tapering to fine point; 3 enlarged tubercles on each side of vent; preanal pores in males 6; preanal groove present; dorsum and forehead reddish-brown to dark grey-black, latter with series of transverse bands, complete or occasionally connected medially to form network of reticulation; dark postocular stripe to armpit; tail with 10 dark bands. DISTRIBUTION Andaman Islands, India. HABITAT AND HABITS Found in lowland rainforests, sometimes entering human habitation. Nocturnal and arboreal. Diet consists of small insects. Clutches comprise 2 eggs.

▪ TRUE GECKOS ▪

Spotted Ground Gecko ▪ *Cyrtodactylus triedrus* 62mm
(*Sinhala* Pulli Vakaniyahuna)

DESCRIPTION Body has small, granular scales, intermixed with larger, keeled scales; midventral scales cycloid and imbricating, numbering 35; toes short; males have 3–4 preanal pores and 3–4 femoral pores; dorsum dark brown to nearly black, with small white spots edged with brown; belly light brown. **DISTRIBUTION** Knuckles Massif, Sri Lanka. **HABITAT AND HABITS** Found in mid-hills, below 700m, occurring under stones and also associated with fallen logs. Diet presumably small arthropods. Clutches comprise 2 eggs, 10 x 12mm. Hatchlings 23mm.

Blotched Ground Gecko ▪ *Cyrtodactylus yakhuna* 41mm
(*Sinhalese* Lapavan Vakaniyahuna, Yak Huna)

DESCRIPTION Body stout, cylindrical, covered with small, granular scales; scales on belly overlapping; tail heavy, tapering and regenerated tail turnip shaped; subcaudal scales not distinctly enlarged; dorsum has 2 cross-rows of dark blotches, consisting of 2 subrectangular marks, and intervening light areas with black spots (subspecies *yakhuna*), or has 2 dark brown transverse rows of large cross-bands, equal to or shorter than light interspaces (subspecies *zonatus*); tail has dark brown bands. **DISTRIBUTION** North East and North Central Provinces (*yakhuna*), and North and North West Provinces (*zonatus*) of Sri Lanka. **HABITAT AND HABITS** Found in deciduous and scrub forests, below 300m. Diet includes arthropods such as termites. Clutches comprise 4 eggs. Incubation period 95 days.

▪ True Geckos ▪

Warty Rock Gecko ▪ *Cyrtopodion kachhense* 43mm
(*Urdu* Kachh Chapkali)

DESCRIPTION Body slender, flattened; head large; snout blunt; eyes large; forehead scales small, with larger tubercles; fingers and toes slender; dorsal tubercles small, and smaller than those on flanks, in longitudinal series, separated by 3–5 rows of smaller granules; blunt spines on tail, comprising lateral rows of scales; males have 4–7 preanal pores; dorsum light brown or grey, with small, irregularly arranged dark spots; belly cream coloured. **DISTRIBUTION** Western India and Pakistan; possibly also Iran. **HABITAT AND HABITS** Found in rocky areas, and enters human habitation. Diet presumably consists of small insects. Clutches comprise 1–2 eggs, 9.5 x 7mm.

Four-clawed Gecko ▪ *Gehyra mutilata* 60mm
(*Sinhalese* Sudhu Geval Huna, Caturanguli Huna)

DESCRIPTION Body stout; head relatively large; skin delicate; tail flattened, widening at base, with sharp, somewhat denticulate edges; large, flat scales on tail and belly; claws on inner digits absent; males have 25–41 preanofemoral pores; dorsum translucent grey to pinkish-grey, with pale vertebral area; indistinct white band along face; belly pale pink. **DISTRIBUTION** Northern and southern India and Sri Lanka; range extends east to southern China, Southeast Asia and Oceania. **HABITAT AND HABITS** Found in forests, but more commonly takes up residence in human habitation. Diet consists of insects. Clutches comprise 2 eggs, 8 x 10.5mm. Hatchlings 17–23mm.

▪ True Geckos ▪

Tokay Gecko ▪ *Gekko gecko* 180mm
(*Assamese* To-khoe; *Bengali* Takshak; *Hindi* Lakri Chipkali; *Manipuri* Chum)

DESCRIPTION Body stout; head large; dorsum scales granular; males have 13–24 preanal pores; dorsum slaty-grey, with red or orange spots; tail dark banded; belly cream coloured, unpatterned or variegated with grey; eye yellow. **DISTRIBUTION** Eastern and north-eastern India, Bangladesh and Nepal; range extends to southern China and Southeast Asia. **HABITAT AND HABITS** Found in lowland forests, and most commonly encountered in villages and towns. Clutches comprise 2 eggs, 25mm. Incubation period 64 days. Hatchlings 40–42mm.

False Bowring's Gecko ▪ *Hemidactylus aquilonius* 50mm
(*Bengali* Choto tiktiki; *Manipuri* Chum)

DESCRIPTION Body slender; head relatively large; snout pointed; males have 12–15 femoral pores; tail-base not swollen, and no denticulate tail edges; dorsum pale brown with darker spots; dark postocular stripe; sometimes, 4 longitudinal streaks along dorsum; tail has dark chevrons; belly unpatterned cream coloured. **DISTRIBUTION** Eastern and north-eastern India, Bangladesh and Nepal; also southern China and Southeast Asia. **HABITAT AND HABITS** Found in human-modified habitats, and frequently seen in houses. Diet consists of insects. Clutches comprise 2 eggs.

▪ TRUE GECKOS ▪

Brooke's House Gecko ▪ *Hemidactylus brookii* 55mm
(*Bengali* Khoskhose Tiktiki; *Dhivehi* Hoanu; *Oriya* Jhitpiti; *Sinhalese* Pulli Geval Huna; *Urdu* Barani Chapkali)

DESCRIPTION Body stout and flattened; head oval; head scales small; dorsal scales granular; rows of tubercles; tail has spine-like tubercles; males have 7–12 preanofemoral pores; dorsum dark brown to light grey, with dark spots arranged in groups; 2 dark lines along nostrils and eyes; belly cream coloured. **DISTRIBUTION** Northern India, Nepal and Pakistan; other extra-limital populations from Borneo, West Africa, southern China and the West Indies need further systematic research. **HABITAT AND HABITS** Found in deserts, parks, gardens and open forests, and enters human habitation. Diet consists of insects. Clutches comprise 2 eggs, 7 x 9mm. Incubation period 43 days.

Depressed Gecko ▪ *Hemidactylus depressus* 80mm
(*Sinhalese* Hali Huna)

DESCRIPTION Body stout; head large, with large granules, especially on snout; midventrals 36–40; digits webbed at base; lamellae under fourth toe 10–11; tail depressed, with serrated lateral edge; males have 16–19 pairs of femoral pores; dorsum light brown or grey, with 4–5 dark transverse, angular markings; dark canthal stripe, edged with a pale one; tail has dark cross-bars; belly greyish-cream. **DISTRIBUTION** Sri Lanka. **HABITAT AND HABITS** Found in lowland areas. Arboreal and rupicolous; associated with trees, boulders and caves, and sometimes enters houses. Diet includes insects. Clutches comprise 2 eggs laid in rock crevices, tree-holes and leaf litter.

■ TRUE GECKOS ■

Yellow-green House Gecko ■ *Hemidactylus flaviviridis* 73mm
(*Bengali* Goda Tiktiki; *Gujarati* Dhedh Garoli, Garoli; *Oriya* Khitpiti; *Rajasthani* Vishamra; *Urdu* Ghar Chapkali)

DESCRIPTION Body stout; head oval; head scales small; dorsum lacks tubercles; tail sometimes has 2 pairs of rows of tubercles; males have fewer than 15 preanaofemoral pores; dorsum pale grey at night to olive by day, when it may show dark cross-bars; belly light yellow. **DISTRIBUTION** Northern and eastern India, Bangladesh and Nepal; range extends west to Red Sea area in northern Africa and Middle East. **HABITAT AND HABITS** Found in light forests and enters human habitation. Diet consists of flies, bugs, mole crickets, beetles, termites, spiders and moths. Clutches comprise 2–3 oval eggs. Incubation period 33–68 days. Hatchlings 11mm.

Asian House Gecko ■ *Hemidactylus frenatus* 67mm
(*Bengali* Mosrin Tiktiki; *Dhivehi* Hoanu; *Gujarati* Garoli; *Hindi* Cheechak, Chipkali, Chiplee; *Kannada* Halli; *Karen* Daw-ly; *Malayalam* Palli; *Marathi* Pal; *Nepali* Mausulee; *Oriya* Jhitpiti; *Sindhi* Chipkali, Chuttee; *Sinhalese* Geval Huna; *Tamil* Gowli, Veettu Palli; *Telugu* Balli; *Urdu* Awara Chapkali)

DESCRIPTION Body slender; head large; dorsal scales smooth; lack of webbing on fingers and toes; sides of tail have enlarged tubercles; no flaps of skin along flanks of body and at

back of hindlimbs; males have 28–36 preanofemoral pores; dorsum greyish-brown, sometimes with darker markings; brown streak with light edge on top runs along sides of head, sometimes continuing along flanks; belly unpatterned cream coloured. **DISTRIBUTION** India (excluding north), Nepal and Sri Lanka; introduced to many parts of tropics and subtropics. **HABITAT AND HABITS** Found in forests as well as human-modified areas. Diet consists of insects and spiders. Clutches comprise 2 eggs, 8 x 10mm.

▪ True Geckos ▪

Garnot's Gecko ▪ *Hemidactylus garnoti* 65mm
(*Manipuri* Chum)

DESCRIPTION Body slender; head large; dorsal scales small; tail depressed, with denticulate lateral edges; 14–19 enlarged femoral scales; dorsum brownish-grey, sometimes with brown and cream spots; belly unpatterned cream coloured. **DISTRIBUTION** North-eastern India, Bangladesh, Bhutan and Nepal; also, South-east Asia and Pacific Ocean islands. **HABITS AND HABITAT** Found in trees as well as walls of buildings, and may enter human habitation. Parenthogenetic; clutches comprise 2 eggs.

Sri Lankan Spotted Rock Gecko ▪ *Hemidactylus hunae* 105mm
(*Sinhalese* Davanta Tit Huna, Kotakka)

DESCRIPTION Body stout; snout somewhat pointed; forehead has large, scattered scales; dorsals not imbricate, with 16–20 rows of tubercles; ventrals smooth; fingers and toes have divided scansors; males have 19–25 femoral pores; mental scales as long as wide; dorsum greyish-brown, with darker, black-edged bands; 2 dark streaks along eye; belly unpatterned cream coloured. **DISTRIBUTION** Eastern Sri Lanka. **HABITAT AND HABITS** Found in low hills and plains of dry and intermediate zones. Associated with rocky outcrops. Diet includes insects and geckos. Clutches comprise 2 eggs, 10 x 13mm. Hatchlings 31mm.

■ True Geckos ■

Sri Lankan Termite Hill Gecko ■ *Hemidactylus lankae* 80mm
(*Sinhalese* Humbas Huna, Kimbula Huna)

DESCRIPTION Body stout; head large; indistinct lateral skin-fold present; forehead and back covered with large, convex tubercles; males have 8 pairs of femoral pores; dorsum

yellowish-olive, with 3–4 large, saddle-like, pale brown patches; brown postocular stripe; belly unpatterned cream; tail dark banded. **DISTRIBUTION** Northern and eastern Sri Lanka. **HABITAT AND HABITS** Found in dry zone, at below 300m. Associated with open forests and scrubland. Nocturnal and terrestrial, occupying rock cracks, tree bark and rodent burrows. Diet includes termites, crickets, grasshoppers, spiders and beetles. Clutches comprise 2–6 eggs, 10 x 12mm.

Bark Gecko ■ *Hemidactylus leschenaultii* 83mm
(*Oriya* Jhitpiti; *Sinhalese* Kabara Huna, Kumbuk Huna, Kimbul Huna, Gas Huna; *Tamil* Maram Palli; *Urdu* Chaal Chapkali)

DESCRIPTION Body stout; head large; tail depressed, its lateral edge spinose; scales small; males have 12–19 femoral pores; dorsum pale grey, with wavy, dark grey or black cross-bars

Juvenile

or series of 'W'-shaped patterns; colouration darker in juveniles than adults; belly unpatterned cream or grey. **DISTRIBUTION** India, Pakistan and Sri Lanka. **HABITAT AND HABITS** Found in lightly wooded areas, and most often seen inside human dwellings as well as on large trees. Diet includes insects and other geckos. Clutches comprise 2 eggs.

TRUE GECKOS

Sri Lankan Spotted Gecko ■ *Hemidactylus parvimaculatus* 55mm
(*Sinhalese* Pulli Geval Huna)

DESCRIPTION Head oval; head scales small; body flattened, with granular scales and rows of tubercles; tail has spine-like tubercles; males have 7–12 preanofemoral pores; dorsum dark brown to light grey, with dark brown spots arranged in groups; 2 dark lines along nostrils and eyes; belly yellowish-cream; tail has indistinct dark bars. **DISTRIBUTION** Sri Lanka, the Maldives and southern India; also Comoros, Mascarene Islands and Seychelles. **HABITAT AND HABITS** Found in parks, gardens and houses, as well as in open forests. Largely terrestrial, though sometimes seen climbing low walls. Nocturnal, and active during hot and wet months. Diet consists of small insects. Clutches comprise 2 eggs, 7 x 9mm. Incubation period about 43 days.

Flat-tailed Gecko ■ *Hemidactylus platyurus* 69mm
(*Manipuri* Chum)

DESCRIPTION Body slender, depressed; snout rather long; fingers and toes about half-webbed; dorsum has granular scales; fringe of skin on sides of body and backs of hindlimbs; 7–9 lamellae under fourth toe; males have 34–36 femoral pores; dorsum light grey, sometimes with darker variegation; dark grey postocular streak reaches shoulder; belly unpatterned cream coloured. **DISTRIBUTION** Eastern and north-eastern India and Andaman and Nicobar Islands, Bangladesh, Bhutan and Nepal; range extends east to eastern China and Southeast Asia. **HABITAT AND HABITS** Found in human habitations and other structures in towns and cities, while being rare in forests, from sea level to at least 1,540m. Diet consists of insects. Clutches comprise 2 eggs, 10.0–10.6 x 8.5–8.9mm. Hatchlings 20.5–25mm.

TRUE GECKOS

Reticulated Gecko ■ *Hemidactylus reticulatus* 40mm

DESCRIPTION Body slender; head relatively short; snout rounded; forehead covered with small granular scales; back has small keeled scales, intermixed with larger, pointed tubercles; tail rounded and covered with pointed tubercles; males have 6–12 preanal pores; dorsum brown, with network of darker lines forming reticulate pattern; belly cream coloured; throat sometimes brown streaked. **DISTRIBUTION** Southern India. **HABITAT AND HABITS** Found in open forests such as scrub country. Nocturnal, hiding under rocks and stones. Diet and reproductive habits unknown.

Termite-hill Gecko ■ *Hemidactylus triedrus* 79mm
(*Urdu* Sahali Chipkali)

DESCRIPTION Body stout; head large; indistinct skin-fold on flanks; forehead and back covered with 16–18 rows of convex tubercules; males have 13–19 preanofemoral pores; dorsum yellowish-olive, with 3 large, saddle-like brown patches, edged with black; yellow postocular stripes to across nape; belly unpatterned cream; tail dark banded. **DISTRIBUTION** Peninsular and western India. **HABITAT AND HABITS** Found in open forests and scrubland. Diet consists of termites, crickets, grasshoppers, spiders and beetles. Clutches comprise 2–6 eggs, 10 x 11.5–12mm. Hatchlings 22mm.

■ True Geckos ■

Western Ghats Worm Gecko ■ *Hemiphyllodactylus aurantiacus* 36mm

DESCRIPTION Body slender; head slightly distinct from neck; broad tail-base; preanal pores 9; femoral pores 7–8 in adult males; dorsum light to mid-brown; 2 series of thin black stripes; dark stripe under eye; distinctly banded tail; belly has orange-red flush, with darker speckling; number of toe scansors 2–3. **DISTRIBUTION** Southern India. **HABITAT AND HABITS** Found in evergreen forests, and sometimes seen on walls of houses. Diet unknown. Clutches comprise 2 eggs, 5mm.

Oceanic Worm Gecko ■ *Hemiphyllodactylus typus* 60mm
(*Sinhalese* Sihin Huna)

DESCRIPTION Body slender; head slightly distinct from neck; broad tail-base; granular dorsal scales; ventral scales smooth, rounded and imbricate; digits free; scansors divided, numbering 3–6; terminal phalange short, clawed; tail prehensile; males have 10–12 angular series of preanal pores, separated from 8–10 femoral pores; dorsum dark brown, with dark brown blotches; dark brown stripe from nostril to shoulder; belly cream coloured, with dark brown speckles.

DISTRIBUTION Andaman Islands, India and Sri Lanka; also southern China, Southeast Asia, east to Oceania and Hawaii. **HABITAT AND HABITS** Found in a variety of habitats, from coastal areas to nearly 1,000m, and may enter human habitation. Diet consists of small insects. Unisexual and parthenogenetic. Clutches comprise 2 eggs, 8 x 6mm. Hatchlings 14.3–17.7mm.

▪ True Geckos ▪

Mourning Gecko ▪ *Lepidodactylus lugubris* 42mm
(*Sinhalese* Salkapa Huna)

DESCRIPTION Body slender; head longer than broad; tubercles on dorsum absent; preanofemoral scales 25–31; dorsum cinnamon-brown or greyish-brown, with brownish-red

tails; dark stripe along head; cross-bars on body and tail 'W'-shaped; belly cream coloured. **DISTRIBUTION** Andaman and Nicobar Islands, India and south-western Sri Lanka; range encompasses large portion of Southeast Asia, east to south-west Pacific. **HABITAT AND HABITS** Found in lowland forests, including mangroves, and associated with leaves and branches. Unisexual and parthenogenetic. Clutches comprise 2 eggs, 7 x 10mm. Hatchlings 17.2–17.8mm.

Andaman Day Gecko ▪ *Phelsuma andamanense* 154mm
(*Bengali* Sabuj Tiktiki; *Hindi* Hara Chipkali; *Karen* Daw-ly La)

DESCRIPTION Body stout; claws reduced and sometimes absent; scales small, granular; eyelid present; pupil rounded; fingers and toes have rounded tips, with undivided scansors

below; males have 15 preanofemoral pores; dorsum of males green with bright red markings, and tail bluish-green; females unpatterned green; belly cream coloured; throat yellow. **DISTRIBUTION** Andaman Islands, India. **HABITAT AND HABITS** Found in lowland forests, including coconut and banana plantations, and occasionally entering human habitation. Clutches comprise 2 oval eggs.

▪ EYELID GECKOS ▪

EUBLEPHARIDAE (EYELID GECKOS)
This family of lizards is related to the true geckos, and includes 36 living species. They lack adhesive toe-pads and have movable eyelids. They lay leathery eggs and the sex of the young is determined by the incubation temperature. Eublepharids include leopard geckos, banded geckos and cat geckos.

Eastern Indian Leopard Gecko ▪ *Eublepharis hardwickii* 110mm
(*Bengali* Kalkut; *Oriya* Kalakuta Shapo)

DESCRIPTION Body stout, flattened; head large; snout bluntly pointed; distinct fleshy eyelids; middorsal tubercles flattened; axillary groove absent; subcaudals not enlarged; dorsum has 2 pale bands between nuchal loop and caudal constriction; pale and dark juvenile banded pattern on dorsum, seen in other members of this genus, is retained.
DISTRIBUTION Eastern Indian peninsula. **HABITAT AND HABITS** Found in dry deciduous forests. Diet consists of insects. Reproductive habits unknown.

Head   *Juvenile*   *Adult*

Common Asian Leopard Gecko ▪ *Eublepharis macularius* 158mm
(*Sindhi* Hun-khun, Khun; *Urdu* Khin-khin, Korrh Kirly)

DESCRIPTION Body stout, flattened; head large; snout bluntly pointed; distinct fleshy eyelids; crescentic ear opening; conical dorsal tubercles; femoral pores 10–14; dorsum of adults pale yellow to brownish-yellow, with small, blue-black spots that may form reticulations; juveniles dark brown to black dorsally, with 2–3 yellow bands across trunk, in addition to white nuchal loop; tail has 4–6 bands. **DISTRIBUTION** Northern India and Pakistan; also Afghanistan. **HABITAT AND HABITS** Found in arid regions such as rocky deserts and scrubland, and associated with rocky crevices, the underneath of rock piles and rat holes. Diet includes crickets, grasshoppers, beetles, dragonflies, antlions, scorpions and lizards. Clutches comprise 1–3 eggs, 13–16 x 31–35mm.

LACERTAS

> **LACERTIDAE (LACERTAS)**
> Lacertas comprise a large (321 described species) family of European wall lizards, the 'true' lizards and grass lizards. They are primarily insectivores. Characteristics include a slender build and long tail, the presence of femoral pores, a large forehead that frequently shows osteoderms, and small, granular scales on the dorsum that are rectangular on the belly.

Indian Fringe-toed Lizard ■ *Acanthodactylus cantoris* 76mm
(*Urdu* Neeli-dum Chalpaya)

DESCRIPTION Body slender; tail long; eyelids movable; lower eyelid translucent; lateral scales small; ventral scales smooth; toes long, fringed; dorsum of adults pink with speckles, reddish-brown or grey with black stripes; juveniles striped yellow and black, with blue tails. **DISTRIBUTION** Northern and northwestern India and Pakistan; also Afghanistan. **HABITAT AND HABITS** Found in landscapes with dry rocky, sandy or alluvial soil, including sea beaches. Excavates shallow burrows at bases of bushes. Diurnal and terrestrial. Diet consists of grasshoppers, beetles, crickets, ants and agamid lizards. Clutches comprise 4–10 eggs.

Sharp-nosed Racerunner ■ *Eremias acutirostris* 70mm

DESCRIPTION Body slender; dorsals granular; flank scales not enlarged; fingers and toes have well-developed fringes; femoral pores 22–30; ventral scales subquadrangular; dorsum pale brown or tan, with fine dark brown speckling forming reticulate pattern.
DISTRIBUTION Western Pakistan; also Afghanistan and Iran. **HABITAT AND HABITS** Found in sand dunes, and inhabits burrows in such environments. Diet includes insects and their larvae. Clutches comprise 2–4 eggs.

■ LACERTAS ■

Snake-eyed Lacerta ■ *Ophisops jerdoni* 41mm
(*Marathi* Tsopai; *Urdu* Khurda Chisma Chalpaya)

DESCRIPTION Body slender; head with large scales; limbs well developed; forehead scales rough; dorsal scales smooth or weakly keeled; femoral pores present in both sexes; fringes on toes absent; dorsum brown or olive, darker on flanks; pale postocular stripe to base of tail; another from upper lip to base of hindlimbs; belly white. **DISTRIBUTION** Western India and Pakistan. **HABITS AND HABITAT** Inhabits dry, rocky terrain. Diet includes grasshoppers, beetles, ants, termites, caterpillars and spiders. Clutches comprise 2–7 eggs, 4–5 x 6.5–7mm. Hatchlings 15–20mm.

Leschenaulti's Lacerta ■ *Ophisops leschenaultii* 50mm

DESCRIPTION Body slender; head has large scales; limbs well developed; anterior lip scales and dorsal scales keeled; tail relatively long; dorsum brown or golden-yellow, with paired black stripe from eye to flanks and tail; second stripe along upper lip to side of body; tail and hindlimbs reddish-brown; forearms brown; belly unpatterned greenish-cream. **DISTRIBUTION** Peninsular India and northern Sri Lanka. **HABITAT AND HABITS** Found in scrub forests and other types of open jungle. Diet consists of insects. Clutches comprise 6 eggs.

LACERTAS/SKINKS

Khasi Hills Long-tailed Lizard ■ *Takydromus khasiensis* 52mm

DESCRIPTION Body slender; head rather long; dorsal surface has large, keeled, plate-like scales; sides of body have large, pointed and keeled scales; tail nearly three times as long as snout-vent length; femoral pores 2–3; dorsum brownish-olive; pale dorsolateral stripe from eye to tail-base, bordered above and below with black spots; black streak along head; belly greenish-cream. **DISTRIBUTION** North-eastern India and Bangladesh; also northern Myanmar. **HABITAT AND HABITS** Found in rocky plateau grassland and adjacent lowland forests. Diet and reproductive habits unknown.

SCINCIDAE (Skinks)
Skinks are members of a large (1,605 described species) family of lizards, found on all continents. Members of the group tend to have relatively small limbs that are often lost altogether, indistinct necks, heavily scaled bodies, a lack of body ornamentation such as crests, and long, tapering tails that can be detached and regrown. Many skink species are fossorial or burrowers. Skinks are typically insectivorous, and reproduce by laying eggs or by giving birth to live young.

SKINKS

Ladakhi Rock Skink ■ *Asymblepharus ladacensis* 55mm
(*Kinnauri* Jalad; *Urdu* Barfani Baahmani)

DESCRIPTION Body slender; 'clear window' in lower eyelid; snout bluntly pointed; limbs have 5 fingers and toes; middorsal scale rows 32–38; tail long, tapering to fine point; dorsum bronze-brown, with dark flecks and scattered, light-edged scales; dark lateral stripe from eye to side of body, enclosing white spots; belly bluish-white. **DISTRIBUTION** Western Himalayas, including northern India, Nepal and Pakistan. **HABITAT AND HABITS** Found in alpine regions at up to 4,200m, in rocky environments such as cliffs, rock walls and edges of rivers. Diet consists of insects. Presumably viviparous.

Four-toed Skink ■ *Chalcidoseps thwaitesii* 107mm
(*Sinhalese* Angili Sathare Heeraluwa)

DESCRIPTION Body slender; snout acute, flattened; eyes small, with scaly lower eyelids; midbody scales smooth, in 24–26 rows; tail elongate, as wide as body; limbs short, 4 toed, inner 2 short; preanal scales enlarged; dorsum dark brown with blackish-brown central region. **DISTRIBUTION** Sri Lanka. **HABITAT AND HABITS** Found in forests at 700–1,000m of Knuckles Massif. Associated with rocks, boulders, decaying logs and leaf litter. Diet consists of insects. Clutches comprise 2 eggs, 6–11 x 18mm.

SKINKS

Nicobarese Tree Skink ■ *Dasia nicobarensis* 98mm

DESCRIPTION Body slender; ear opening small; snout pointed; scales under tail enlarged; dorsum olive-brown with pale stripe along flanks and tail-base; belly bluish-cream, each scale pale centred.

DISTRIBUTION Car Nicobar and Great Nicobar, India. HABITAT AND HABITS Found in lowland forests, and sometimes enters human settlements. Diurnal and arboreal. Diet presumably comprises arthropods. Reproductive habits unknown.

Andaman Grass Skink ■ *Eutropis andamanensis* 105mm

DESCRIPTION Body stout; head nearly indistinct from neck; lower eyelids scaly; dorsal scales have 5–7 large keels; dorsum brown, with 2 series of black spots along vertebral line; dark postocular stripe runs to front half of flanks; belly yellowish-cream; sides of head, neck and belly bright red in breeding season. DISTRIBUTION Andaman Islands of India, in addition to Cocos group of Myanmar. HABITAT AND HABITS Found in lowland rainforests. Diurnal and terrestrial. Diet includes insects. Reproductive biology unknown.

SKINKS

Keeled Grass Skink ■ *Eutropis carinata* 125mm
(*Bengali* Anchil, Anjon, Telpa Shap; *Gujarati* Arooni-mashi, Samp-ni-mashi, Samp Saradi; *Hindi* Bhabani, Loten; *Kannada* Arani; *Malayalam* Arana; *Marathi* Sapswulli, Sarpa Chi Mausi; *Nepali* Pate Telia, Vanmungree; *Oriya* Champei Neula; *Rajasthani* Nagarbamni; *Sinhalese* Garendi Hickanalla, Thel Hickanalla; *Tamil* Pambarannai, Periya Arene; *Telugu* Nallikalla Pamu)

DESCRIPTION Body stout; head indistinct from neck; limbs well developed; lower eyelids scaly; dorsal scales have 3–8 keels; ventral scales smooth; dorsum bronze-brown or olive, with yellow lateral band; broad chocolate-brown band on top; belly cream or yellow, darkening posteriorly. During breeding season, flanks of males are brick-red.

DISTRIBUTION India (except north-eastern region), Bangladesh, Nepal and Sri Lanka. **HABITAT AND HABITS** Found in many habitat types, from forests and edges of deserts, to scrubland, parks and gardens, and even urban areas. Diurnal and terrestrial. Diet consists of arthropods and other invertebrates, including crickets, caterpillars, beetles and earthworms, in addition to small vertebrates. Clutches comprise 2–8 eggs, 10–12 x 16–17mm. Incubation period more than 30 days. Hatchlings 12–12.5mm.

Striped Grass Skink ■ *Eutropis dissimilis* 150mm
(*Bengali* Anjon; *Punjabi* Pairon-wala Samp; *Urdu* Ghas Guddi)

DESCRIPTION Body stout; snout short; lower eyelid has 'clear window; scales have 3 keels; ventral scales smooth; dorsum dark brown or pale olive, with yellow stripes, edged with black dots, sometimes fused to form lines; flanks have small white spots; pale subocular stripe; belly greenish-white.

DISTRIBUTION Northern and north-eastern India, Nepal and Pakistan; also Afghanistan. **HABITAT AND HABITS** Found in damp grassland. Diurnal and terrestrial, and associated with holes of mole rats along river banks. Diet consists of insect and spiders; frogs also reportedly taken. Clutches comprise 6 eggs.

SKINKS

Bronze Grass Skink ■ *Eutropis macularia* 75mm
(*Manipuri* Charu Chum; *Urdu* Bhoori Gaas-guddi)

DESCRIPTION Body slender; limbs well developed; dorsal scales have 5–9 keels; lower eyelid lacks 'clear window'; dorsum bronze-brown, with or without spots, depending on locality; flanks darker, spotted with white, especially in juveniles and males, and brown or grey in females; belly cream coloured; breeding males have bright red lips and flanks. **DISTRIBUTION** India, Bangladesh, Bhutan, Nepal, Pakistan and Sri Lanka; also Southeast Asia. **HABITAT AND HABITS** Found in deciduous to evergreen forests, as well as plantations. Diet includes beetles and grasshoppers. Clutches comprise 2–4 eggs, 13–15 x 7–8mm.

Many-lined Grass Skink ■ *Eutropis multifasciata* 137mm
(*Manipuri* Charu Chum)

DESCRIPTION Body stout; lower eyelid scaly; dorsal scales have 3, rarely 5, keels; dorsum bronze-brown, usually with dark brown band, and series of pale spots or streaks along flanks; breeding males have bright orange or reddish-orange band on flanks; pale dorsolateral line; belly cream coloured. **DISTRIBUTION** North-eastern India and the Nicobar Islands; also Southeast Asia and southern China. **HABITAT AND HABITS** Found in forest edges and open forests, as well as disturbed habitats such as clearings around human settlements. Diurnal and terrestrial. Diet consists of insects. Ovoviviparous, giving birth to 2–10 live young, 36–43mm.

▪ SKINKS ▪

Four-keeled Grass Skink ▪ *Eutropis quadricarinata* 50mm
(*Manipuri* Charu Chum)

DESCRIPTION Body stout; head indistinct from neck; lower eyelid scaly; dorsal scales have 4 sharp keels; dorsum olive-brown, unpatterned or with small black spots arranged longitudinally; flanks sometimes have dark lines; belly and upper lip cream coloured. **DISTRIBUTION** North-eastern India; also Myanmar. **HABITAT AND HABITS** Found in evergreen forest edges and clearings. Diurnal and terrestrial. Dietary habits unknown. Clutches comprise 3 eggs, 10.6–10.8 x 6.3–6.5mm.

Tammenna Skink ▪ *Eutropis tamanna* 57mm
(*Sinhalese* Tambapanni Hikanala)

DESCRIPTION Body stout; snout short; no transparent disks on lower eyelids; dorsal scales have 4–5 keels; dorsum medium brown; lips bright orange in males, yellow in females, extending to midflanks; black stripe from below eye to beyond base of tail, with large yellow-cream spots – stripe paler in females than males; belly yellowish-cream. **DISTRIBUTION** Northern Sri Lanka. **HABITAT AND HABITS** Found in lowlands close to sea beaches, under heaps of debris such as coconut husk piles. Diurnal and terrestrial. Diet includes insects. Clutches comprise 2 eggs.

Male

Head of female

Head of male

Female

SKINKS

Tytler's Grass Skink ■ *Eutropis tytlerii* 150mm

DESCRIPTION Body stout; head large; adults have swollen cheeks; lower eyelid scaly; dorsal scales have 3 weak keels, central one least well defined; dorsum bronze-brown,

unpatterned or with darker spots; belly greenish-cream. DISTRIBUTION Andaman Islands, India. HABITAT AND HABITS Found in lowland rainforests as well as open forest edges. Diurnal and terrestrial. Diet includes insects and other invertebrates. Reproductive habits unknown.

Deignan's Lanka Skink ■ *Lankascincus deignani* 58mm
(*Sinhalese* Deignange Lakhiraluva)

DESCRIPTION Body stout; midbody scale rows 24–28; dorsum olive-brown; thick, dark lateral stripe, edged above by brownish-yellow stripe, and below by 3–4 grey stripes extends from edge of eye to tail-tip; belly cream-white or pale pink; black spots on upper jaw. DISTRIBUTION Central Sri Lanka. HABITAT AND HABITS Found in forested mid-hills to submontane areas, at 600–1,700m. Associated with moist leaf litter, living under stones and logs. Diet includes insects. Clutches comprise 2 eggs.

▪ Skinks ▪

Deraniyagala's Lanka Skink ▪ *Lankascincus deraniyagalae* 43mm
(*Sinhalese* Daraniyagalge Lakhiraluva)

DESCRIPTION Body slender; tail more than 1.5 times size of body; midbody scale rows 26–28; dorsum medium brown; dark brown stripe from eye to hindlimbs laterally, extending to middle of tail; scattered white spots laterally; throat blue with pale spots; belly and tail yellow. **DISTRIBUTION** Central Sri Lanka. **HABITAT AND HABITS** Found in forested mid-hills, at 700–1,000m. Associated with moist leaf litter, living under stones and logs. Diet includes insects. Clutches comprise a single egg.

Common Lanka Skink ▪ *Lankascincus fallax* 43mm
(*Sinhalese* Sulaba Lakhiraluva)

DESCRIPTION Body slender; head indistinct from body; frontoparietal fused; midbody scale rows 24–28; lamellae under fourth toe 13–18; dorsum pale to dark brown; dorsal scale has pale stripe joining to form longitudinal line on dorsum; yellowish-brown stripe from posterior edge of eye to beyond mid-tail; throat colour differs considerably, being red, blue or cream, with white spots; belly cream; ventral scales have frosted pattern, forming longitudinal lines. **DISTRIBUTION** Sri Lanka. **HABITAT AND HABITS** Found in many habitat types, from sea level to about 1,050m. Diurnal and semifossorial. Diet includes insects. Reproductive habits unknown.

▪ Skinks ▪

Gans's Lanka Skink ▪ *Lankascincus gansi* 40mm
(*Sinhalese* Gansge Lakhiraluva)

DESCRIPTION Body slender; head indistinct from body; midbody scale rows 23–28; lamellae under fourth toe 12–16; dorsum greyish-brown, with brownish-black vertebral and

flank-stripes; flanks spotted with yellowish-cream; iris of eye yellowish-brown; throat of males dark and belly unpatterned greyish-yellow. **DISTRIBUTION** Central and south-western Sri Lanka. **HABITAT AND HABITS** Found in forests at 30–700m in wet climatic zone; enters gardens, secreting itself under stones, in leaf litter and under logs. Diet consists of insects. Clutches comprise a single egg.

Taylor's Lanka Skink ▪ *Lankascincus taylori* 52mm
(*Sinhalese* Telorge Lakhiraluva)

DESCRIPTION Body slender; head small; tail long; midbody scale rows 24–26; supralabials 7; lamellae under fourth toe 12–18; dorsum chocolate-brown, each scale with dark grey horseshoe mark, open-ended posteriorly; dark brown flank-band with blue spots; throat grey with blue spots; belly yellow. **DISTRIBUTION** Central Sri Lanka. **HABITAT AND HABITS** Found in montane forests at 470–1,350m. Associated with leaf litter, stones and logs. Diet consists of insects. Clutches comprise 2 eggs.

▪ Skinks ▪

Small-eared Striped Skink ▪ *Lipinia macrotympana* 38mm

DESCRIPTION Body slender; head small, distinct from neck; transparent disk on lower eyelid; snout acute; limbs rather small; tail elongate, rounded, ending in sharp point; tympanum exposed, borders lacking lobules; dorsal scales smooth; ventral scales overlapping; dorsum has yellow vertebral stripe; paravertebral region has grey stripes; belly unpatterned cream; tail bright orange. **DISTRIBUTION** Andaman and Nicobar Islands, India. **HABITAT AND HABITS** Found on sea beaches bordering lowland rainforests. Diet unknown. Clutches comprise 2 eggs.

White-spotted Supple Skink ▪ *Lygosoma albopunctata* 60mm
(*Dhivehi* Gahaheta)

DESCRIPTION Body elongate; head indistinct from neck; lower eyelid scaly; ear opening rounded; scales smooth or feebly keeled; tail thick, rounded, tapering to narrow point; dorsum brown to reddish-brown; each scale with dark spot forming longitudinal series; sides of neck and flanks dark brown or black, spotted with white; belly unpatterned yellowish-white; tail of juveniles bright red, turning brown with growth. **DISTRIBUTION** Eastern and north-eastern India, and Nepal. **HABITAT AND HABITS** Found in hills and plains, and associated with rocky substrate. Diet comprises small insects. Reproductive habits unknown.

▪ SKINKS ▪

Bowring's Supple Skink ▪ *Lygosoma bowringii* 56mm

DESCRIPTION Body slender, elongate; head indistinct from neck; lower eyelid scaly; ear opening rounded; scales smooth or weakly keeled; tail thick, rounded, tapering to narrow point; dorsum bronze-brown; flanks have dark band, with white and black spots; belly yellow; tail of juveniles bright red, turning grey or brown with growth. DISTRIBUTION Andaman and Nicobar Islands, India; also Southeast Asia and eastern China. HABITAT AND HABITS Found in relatively open areas such as human habitation. Diurnal and subfossorial. Diet consists of small insects. Clutches comprise 2–4 eggs.

Spotted Supple Skink ▪ *Lygosoma punctata* 85mm
(*Bengali* Shoru Anjan; *Gujarati* Samp Saradi; *Urdu* Chitri Baghban Baamani)

DESCRIPTION Body elongate; head indistinct from neck; lower eyelid has transparent disk; ear opening rounded; scales smooth; tail rather thick, rounded, tapering to narrow point; dorsum bronze-brown, with 4–6 rows of black spots, lateral spots more distinct; broad cream flank-stripe; belly unpatterned cream; tail of juveniles bright red, turning brown or pink with growth. DISTRIBUTION India, Bangladesh, Nepal, Pakistan and Sri Lanka. HABITAT AND HABITS Found in hills and plains. Associated with leaf litter, and frequently enters houses. Diurnal and subfossorial. Diet consists of small insects. Clutches comprise 2–4 eggs.

▪ Skinks ▪

Smith's Snake Skink ▪ *Nessia bipes* 80mm
(*Sinhalese* Smithge Sarapa Heeraluwa)

DESCRIPTION Body slender, of equal girth from head to tail; snout broad and obtuse; forelimbs absent; hindlimbs bud-like; midbody scale rows 28; dorsum brown or light reddish-brown; hatchlings dark grey or black. **DISTRIBUTION** Eastern Knuckles Massif, Sri Lanka. **HABITAT AND HABITS** Found in submontane forests at up to 750m. Associated with decaying logs, loose soil and leaf litter. Diet includes insects and earthworms. Clutches comprise 2–4 eggs.

Toeless Snake Skink ▪ *Nessia monodactylus* 90mm
(*Sinhalese* Ananguli Sarapa Heeraluwa)

DESCRIPTION Body slender, of equal girth from head to tail; snout short and obtuse; lower eyelid has transparent disk; limbs lack digits; midbody scale rows 26–28; dorsum brown, dark grey or black. **DISTRIBUTION** Central Highlands and associated lowlands of Sri Lanka. **HABITAT AND HABITS** Found in forests in wet climatic zone at 400–1,500m. Associated with decaying logs, loose soil and humus. Diet includes insects. Clutches comprise 2 eggs.

SKINKS

Indian Sandfish ■ *Ophiomorus raithmai* 95mm
(*Punjabi* Reg-mahi; *Sindhi* Raith Mai; *Urdu* Reg-mahi)

DESCRIPTION Body slender; head indistinct from neck; lower eyelid has 'clear window'; eyes reduced; ear opening tiny; fingers 3; toes 3; third toe has 4–6 scales; dorsum pale

brown or yellowish-brown; belly paler.
DISTRIBUTION Western India and Pakistan.
HABITAT AND HABITS Found in sandy deserts, and abundant in inter-dune regions. Fossorial to surface living, dwelling at bases of shrubs. Diet includes locusts, as well as termites, moths, grasshoppers, cockroaches and beetles. Reproductive habits unknown.

Dussumier's Litter Skink ■ *Sphenomorphus dussumieri* 60mm

DESCRIPTION Body slender; head distinct from neck; snout short; tympanum situated on surface; dorsal scales smooth, with fine striation; limbs relatively short; dorsum bronze-

brown with dark spots; pale postocular stripe extends to flanks as dorsolateral stripe, inner edge with dark brown, white-spotted streak; belly cream coloured.
DISTRIBUTION Western Ghats, south-western India. **HABITAT AND HABITS** Found in evergreen and moist deciduous forests, and plantation forests such as rubber plantations, in mid-hills. Diurnal and terrestrial. Diet and reproductive habits unknown.

SKINKS

Himalayan Litter Skink ■ *Sphenomorphus indicus* 90mm
(*Nepali* Vanmungree)

DESCRIPTION Body slender; head distinct from neck; snout short; tympanum deep; dorsal scales smooth; dorsum brown, unpatterned or with dark brown spots forming longitudinal lines; lips dark barred; belly cream coloured.
DISTRIBUTION Eastern and north-eastern India, Bangladesh, Bhutan and Nepal; also southern China and Southeast Asia.
HABITAT AND HABITS Found in lowland evergreen forests. Diurnal and semifossorial, hiding in leaf litter. Diet includes insects. Ovoviviparous, producing 4–11 young.

Spotted Litter Skink ■ *Sphenomorphus maculatus* 62mm

DESCRIPTION Body slender; head distinct from neck; snout short; tympanum on surface, not deeply sunk; dorsal scales smooth; dorsum bronze-brown or brownish-pink, unpatterned or with dark green spots; 2 dark median series of spots; dark flank-stripe, spotted with white; belly cream, turning yellow during breeding season. **DISTRIBUTION** Eastern and north-eastern India, Andaman and Nicobar Islands, Bangladesh and Nepal; also southern China and Southeast Asia.
HABITAT AND HABITS Found in plains, close to streams including on seashores and at edges of mangrove swamps, to mid-hills. Diurnal and terrestrial. Diet comprises spiders, crickets and moths. Oviparous, producing 4–5 eggs.

◾ Skinks/Glass Snakes ◾

North-eastern Water Skink ◾ *Tropidophorus assamensis* 67mm

DESCRIPTION Body slender; head rather small; snout pointed; tympanum at surface; lower eyelid scaly; preanal scales enlarged; dorsum brown, with light and dark markings; 2

broad, dark-edged yellow dorsal cross-bars, one across shoulders, one across tail-base; tail dark barred; belly light brown with dark longitudinal steaks. **DISTRIBUTION** North-eastern India and Bangladesh. **HABITAT AND HABITS** Found at edges of small, rocky streams in low hills. Diet comprises insects. Reproductive habits unknown; likely to be ovoviviparous.

ANGUIDAE (Glass Snakes)
The Anguidae is a large and diverse family of 77 species including slow worms and glass lizards. Anguids are carnivorous or insectivorous, and include both egg-laying and viviparous species. They have hard osteoderms under the scales, an armour plate-like appearance to the dorsum and chisel-like teeth.

Eastern Glass Snake ◾ *Dopasia gracilis* 18cm
(*Khasia* Naingbaen; *Tangkhul* Hingkorfun)

DESCRIPTION Body slender; limbs absent; ear openings as large as nostrils, nearly circular; lateral fold present; middorsals keeled; dorsum pale brown to brick-red, with dark

brown flank-stripe; vertebral spots sometimes present; transverse series of sky-blue spots during breeding season; belly pale brown or yellow. **DISTRIBUTION** Northern to north-eastern India and Nepal; also Southeast Asia. **HABITAT AND HABITS** Found in submontane and montane forests at 900–2,400m. Diurnal and terrestrial. Diet consists of arthropods. Clutches comprise 4–7 eggs, 12 x 18mm.

WORM LIZARDS/MONITORS

DIBAMIDAE (WORM LIZARDS)
The worm lizard family comprises 23 described species. They are small to medium in size (with a snout to vent length of 80–200mm), fossorial and insectivorous.

Nicobarese Worm Lizard ■ *Dibamus nicobaricum* 135mm

DESCRIPTION Body slender, elongate; snout conical; head distinct from neck; forehead covered with enlarged scales; forelimbs and hindlimbs absent in females; in males, hindlimbs reduced to flaps; tail short; dorsal scales smooth; dorsum iridescent chestnut, with pale cream neck and throat-band; darker cream midbody band grades into chestnut at back; belly pink; pale spot over vent. **DISTRIBUTION** Kamorta and Great Nicobar Island, India. **HABITAT AND HABITS** Found in alluvial soil in rainforests. Diet unknown, and likely to consist of insect pupae and earthworms. Reproductive habits unknown.

VARANIDAE (MONITORS)
Monitors belong to a fairly small (79 described species) family of lizards that includes some of the largest species. While predominantly carnivorous or scavengers, a few are known to be frugivores. Most are active predators of small to medium-sized vertebrates, chasing down mammalian prey.

Bengal Monitor ■ *Varanus bengalensis* 1.7m

(*Assamese* Gui Shap; *Bengali* Go-shap, *Gujarati* Chandan Goh; *Hindi* Goh; *Kannada* Udumbu; *Malayalam* Udumbu; *Manipuri* Hangkok; *Marathi* Ghorpad; *Nepali* Suraj Goharo; *Oriya* Godhi; *Punjabi* Gar-gho; *Pushtu* Somsereh; *Rajasthani* Chandan Goh; *Sindhi* Kala Goh; *Sinhalese* Thala Goya; *Tamil* Udumbu; *Telugu* Udumu; *Urdu* Goh)

DESCRIPTION Body slender; snout somewhat elongate; nostrils nearer eye than snout-tip; nostril an oblique slit; nuchal scales rounded; crown scales larger than nuchal scales; midventral scales smooth; tail flattened; juveniles pale or dark dorsally, with yellow bands of spots in transverse series; snout unpatterned; belly cream or yellow, lacking dark vertical 'V'-shaped marks extending to sides of belly. **DISTRIBUTION** India, Bangladesh, Bhutan, Nepal, Pakistan and Sri Lanka; also Afghanistan and Myanmar. **HABITAT AND HABITS** Found in semi-deserts and scrub, to evergreen forests and plantations. Diet consists of insects, spiders, snails, crabs, frogs, mammals, birds, lizards and snakes, as well as carrion. Clutches comprise 12 eggs, 29 x 15mm. Hatchlings 94mm.

◾ Monitors ◾

Yellow Monitor ◾ *Varanus flavescens* 83cm
(*Assamese* Gui Shap, Irong; *Bengali* Go-shap, Hungui Shap, Shorno Godhika, Shona gui; *Hindi* Dabkharabbi Goh, Vishkhampar; *Nepali* Bhainse Goharo, Cheparo, So Samp; *Oriya* Godhi; *Punjabi* Peeli Goh; *Urdu* Peeli Goh)

DESCRIPTION Body stout; snout short and convex; nostril an oblique slit, closer to snout-tip than to orbit; nuchal scales strongly keeled; crown scales smaller than nuchals; midventral scales smooth; tail strongly compressed; juveniles have transverse rows of fused yellow spots on dark background; dorsum light to dark brown, with brownish-red areas between yellow transverse bands.

DISTRIBUTION India, Bangladesh, Nepal and Pakistan. **HABITAT AND HABITS** Found in wetlands such as marshes, as well as flooded fields of rice paddy. Diet consists of insects, earthworms, amphibian and reptile eggs, birds and their eggs, and rats. Clutches comprise 4–30 eggs, 37 x 21mm. Incubation period 149–155 days.

Water Monitor ◾ *Varanus salvator* 3.2m
(*Bengali* Boro Gui, Kalo Gui, Ram Godhika; *Car Nicobarese* Kao; *Central Nicobarese* Huye; *Karen* Trey; *Manipuri* Hangkok; *Oriya* Panigodhi; *Sinhalese* Kabara Goya; *South Nicobarese* Okri, Ukaine)

DESCRIPTION Body stout; snout depressed; nostril rounded or oval, twice as far from orbit as from snout-tip; nuchal scales strongly keeled; crown scales larger than nuchal scales; midventral scales feebly keeled; tail strongly compressed with double-toothed crest above; juveniles dark dorsally, yellow spotted or occelli in transverse series; snout black barred; belly yellow with narrow, vertical black 'V'-shaped marks extending to flanks. **DISTRIBUTION** Eastern and north-eastern India; Andaman and Nicobar Islands, Bangladesh, Bhutan and Sri Lanka; also southern and eastern China, and Southeast Asia.

HABITAT AND HABITS Found in wetlands such as marshes, estuaries, rivers and canals in cities, mangroves and streams in dipterocarp forests. Diet

includes insects, fish, crabs and turtles, as well as eggs of water birds, crocodiles and turtles; also known to scavenge. Clutches comprise 7–30 eggs, 32–43 x 67–83mm.

◾ WART SNAKES/PIPE SNAKES ◾

ACROCHORDIDAE (WART SNAKES)
The three known species of this aquatic snake family have dorsally placed eyes, wrinkly or baggy skin, and tiny, wart-like scales on their bodies. They are ambush feeders, the rough scales evidently permitting them to hold onto slippery fish.

Wart Snake ◾ *Acrochordus granulatus* 100cm
(*Bengali* Anchil Shap, Reti; *Gujarati* Halero; *Sinhalese* Diya Goya, Mada Panuva, Redi Naya)

DESCRIPTION Heavy body; head short and blunt, covered with granular scales; nostrils valve-like, situated on top of snout; tail laterally compressed, ending in point; skin has numerous small scales; head dark grey or black; body has alternating grey and cream-coloured bands, tapering towards belly.
DISTRIBUTION Coasts of India, Bangladesh and Sri Lanka; also Southeast Asia. **HABITAT AND HABITS** Found in coastal and mangrove areas, and able to remain submerged underwater for more than 2 hours. Nocturnal; associated with mudflats and shallow coastal waters, sometimes entering rivers. Diet includes fish such as gobies and their relatives, and crustaceans. Ovoviviparous; clutches comprise 4–12 young, measuring 23cm.

CYLINDROPHIIDAE (PIPE SNAKES)
The Cylindrophiidae family includes 14 species of non-venomous 'pipe snake'. They are burrowing species with a number of adaptations for fossoriality, such as a small head that is indistinct from the neck, strongly joined cranial bones, stout, cylindrical bodies, small eyes, short and blunt-tipped tails, and moderately long, stout teeth.

Sri Lankan Pipe Snake ◾ *Cylindrophis maculata* 715cm
(*Sinhalese* Depath Naya, Wataulla)

DESCRIPTION Body stout; head depressed; snout broadly rounded; neck indistinct; eyes small; tail short; dorsal scales smooth, shiny and iridescent; dorsum brick-red with black network enclosing 2 spots; belly cream coloured, with black bars or variegated pattern. **DISTRIBUTION** Sri Lanka. **HABITAT AND HABITS** Found in open forests and more disturbed areas, from plains to about 1,000m. Subfossorial, hiding under stones, logs and rocks. Diet includes snakes, earthworms and insects. Ovoviviparous, producing 1–15 live young, 105–191mm.

◾ SHIELDTAILED SNAKES ◾

> **UROPELTIDAE (SHIELDTAILED SNAKES)**
> There are 55 described species in the Uropeltidae family. These small snakes can be recognized by the distinct keratinous shield at the tip of the tail, a primitive, inflexible skull, vestigial eyes, smooth, iridescent scales and small ventral scales.

Blyth's Shieldtail ◾ *Rhinophis blythii* 37cm
(*Sinhalese* Gomarathudulla)

DESCRIPTION Body slender; head small; snout acute; neck indistinct; tail short, ending in shield with minute spines; dorsals smooth, and iridescent under sunlight; ground colour brownish-black with yellow markings; tail region has yellow area enclosing black patch dorsally. **DISTRIBUTION** Sri Lanka. **HABITAT AND HABITS** Found in forested mid-hills, as well as tea estates. Lives in humus and leaf litter, and sometimes among roots of plantain, in silted drains and in vegetable beds. Diet comprises earthworms. Ovoviviparous, producing 3–6 young.

Drummond-Hay's Shieldtail ◾ *Rhinophis drummondhayi* 33cm

DESCRIPTION Body slender; snout acute; rostral lacks ridge; dorsum brown; scales have a yellow margin; belly dirty yellow, each scale with brown blotch; partial or complete ring at tail-base. **DISTRIBUTION** Sri Lanka. **HABITAT AND HABITS** Found in forests in mid-hills, at up to 1,200m, and also encountered close to plantations. Associated with silted drains within tea estates, areas under decaying vegetation and logs, and humus near cattle sheds. Diet comprises earthworms. Ovoviviparous, producing 2–5 young.

■ Shieldtailed Snakes ■

Hemprich's Shieldtail ■ *Rhinophis homolepis* 28cm
(*Sinhalese* Depaththudulla)

DESCRIPTION Snout acute; caudal disk convex; dorsum black-blue, each scale yellow margined; flanks have yellow spots; belly paler than dorsum, with large yellow scale margins.

DISTRIBUTION Central Sri Lanka.

HABITAT AND HABITS Found in submontane forests and disturbed areas. Associated with loose soil in forest edges and agricultural land, as well as cattle sheds. Diet includes earthworms. Clutches comprise 2–4 young, 90mm.

Ashambu Shieldtail ■ *Uropeltis liura* 32cm

DESCRIPTION Body slender; snout acutely pointed; dorsals smooth; dorsum purplish-brown; each scale dark edged; transverse series of small, black-edged yellow occelli; belly has large, alternating black-and-yellow spots or cross-bars.

DISTRIBUTION Western Ghats of south-western India.

HABITAT AND HABITS Found as in hilly forests at 135–1,500m, as well as in plantations and gardens. Fossorial, burrowing to depths of 30cm in gardens as well as edges of cardamom plantations. Ovoviviparous, producing 4 young.

▪ Shieldtailed Snakes/Pythons ▪

Large-scaled Shieldtail ▪ *Uropeltis macrolepis* 30cm

DESCRIPTION Body slender; snout rounded; tip of tail truncated obliquely; tail-shield large, with small spines; dorsum black or dark purplish-brown, scales light margined; short yellow or orange stripe on lips and sides of neck; broad yellow or orange stripe along tail; belly black or purplish-brown. DISTRIBUTION Northern Western Ghats of southwestern India. HABITAT AND HABITS Found in hilly forests. Fossorial. Diet presumably comprises earthworms and insects. Reproductive habits unknown.

PYTHONIDAE (PYTHONS)
Pythons are a group of non-venomous snakes comprising 40 named species. Most species are ambush predators of warm-blooded prey, using constriction (as do boas) to kill large prey via asphyxiation.

Reticulated Python ▪ *Malayopython reticulatus* 9.80m
(*Bengali* Ajogar, Golbahar; *Car Nicobarese* Tulan; *Central Nicobarese* Tulan; *Shompen* Bhungei; *South Nicobarese* Tolat, Yama)

DESCRIPTION Body long, relatively slender; sensory pits in rostral and first 4 supralabials and some infralabials present; dorsum yellow or light brown, with series of dark brown spots, each edged with black; dark streak along forehead and another on each side of head. DISTRIBUTION Nicobar Islands, India and eastern Bangladesh; also Southeast Asia. HABITAT AND HABITS Found in lowland forests in the region. Diet comprises birds and mammals. Oviparous, producing clutches of 15–100 eggs, which are incubated by the mother and hatch in 65–105 days.

◾ PYTHONS ◾

Burmese Rock Python ◾ *Python bivittatus* 7m
(*Assamese* Ajagar, Jalpehia; *Bengali* Ajogar, Kalo-matha Moyal, Ulubora; *Manipuri* Lairen; *Nepali* Anginger)

DESCRIPTION Body stout; head lance shaped; sensory pits in rostral and first 2 supralabials and some infralabials present; suboculars separate supralabials from eye; spurs small; dorsum dark brownish-grey above, with series of large, squarish dark grey or brown marks; rounded or irregular-shaped blotches on flanks lack light centres; dorsal and lateral spots are darker; dark subocular stripe. **DISTRIBUTION** Eastern and north-eastern India, eastern Bangladesh, Bhutan and Nepal; also mainland Southeast Asia. **HABITAT AND HABITS** Found in evergreen and deciduous forests, grassland and mangroves, from lowlands, up to mid-hills and rarely above 1,500m. Diet includes warm-blooded prey, such as deer and goat. Clutch size up to 100 eggs, which are guarded by the mother.

Indian Rock Python ◾ *Python molurus* 7.6m
(*Bengali* Ajogar, Lal-matha Moyal; *Gujarati* Ajagar; *Hindi* Ajgar; *Kannada* Hebau Havu; *Malayalam* Malam Pambu, Perum Pambu; *Nepali* Anginger; *Oriya* Ajagara; *Punjabi* Ayydha; *Sindhi* Azdar; *Sinhalese* Dhara Pimbura; *Tamil* Kaloodai Viriyan, Malai Pambu, Peria Pambu; *Telugu* Konda Chiluva)

DESCRIPTION Body stout; head lance shaped; sensory pits in rostral and first 2 supralabials and some infralabials present; sixth or seventh supralabial contacts eye; spurs small; dorsum yellowish-grey to mid-brown above, with series of large, squarish dark grey or brown marks; flanks have rounded or irregular-shaped blotches with light centres. **DISTRIBUTION** Peninsular India, western Bangladesh, Nepal, eastern Pakistan and Sri Lanka. **HABITAT AND HABITS** Associated with forests and scrubland. Emerges from burrows of porcupines, bears and rodents to ambush warm-blooded prey and monitor lizards. Oviparous, producing clutches of 8–107 eggs, 120 x 60mm, which are laid in damp soil and guarded by the mother. Incubation period 58–72 days.

◾ SUNBEAM SNAKES/BOAS ◾

XENOPELTIDAE (SUNBEAM SNAKES)
The Xenopeltidae are referred to as sunbeam snakes due to their highly iridescent scales. There are two known species, which inhabit burrows. They have large head scales; the ventral scales are slightly reduced and pelvic vestiges are absent. Sunbeam snakes kill their prey by constriction, in a similar way to their python relatives.

Sunbeam Snake ◾ *Xenopeltis unicolor* 125cm
(*Hindi* Tael Samp)

DESCRIPTION Body stout; head indistinct from body, snout depressed, rounded; eyes small, pupils vertical; nostrils situated between 2 nasals; tail short; dorsum iridescent

brown, each scale light edged; belly white or cream coloured; juveniles have pale yellow or cream-coloured collar. **DISTRIBUTION** Nicobar Islands, India; also south-eastern China and Southeast Asia. **HABITAT AND HABITS** Found in swamps and lowland forests. Nocturnal and semifossorial. Diet includes rodents, ground-nesting birds, amphibians and snakes. Clutches comprise 6–17 eggs, 18 x 58mm.

BOIDAE (BOAS)
Boas belong to a small group (61 living species) of primitive non-venomous snakes. There are several differences between boas and pythons, including a lack in the former group of postfrontal bones or premaxillary teeth, as well as in showing ovoviviparity (pythons produce eggs). Boas catch warm-blooded prey such as birds and mammals by ambush.

Common Sand Boa ◾ *Eryx conicus* 100cm
(*Bengali* Bali Bora, Thuthuk; *Gujarati* Gadhiya Samp; *Hindi* Do-muha; *Malayalam* Mano-lage Pambu; *Marathi* Durkiya Ghonas; *Oriya* Boda Sapa; *Punjabi* Do-muhi Sapp; *Sinhalese* Vali Pimbura; *Tamil* Mann Pudeyan; *Telugu* Erra Poda Pamu; *Urdu* Do-muhi Samp)

DESCRIPTION Body short, stout, cylindrical; dorsals keeled, head indistinct from neck; mental groove absent; snout rounded; nostrils and eyes small; pair of spurs on each side of vent; tail short, tapering to acute point; tail scales strongly keeled; dorsum brownish-grey,

with series of large, irregular dark brown or reddish-brown blotches, sometimes fused or forming zigzag pattern. **DISTRIBUTION** Western and peninsular India, Nepal, Pakistan and Sri Lanka; unconfirmed records from Bangladesh. **HABITAT AND HABITS** Found in arid regions with loose, sometimes sandy soil. Active at dusk. Diet includes small mammals and birds. Ovoviviparous; clutches comprise 3–16 young.

▪ BOAS ▪

Red Sand Boa ▪ *Eryx johnii* 125cm
(*Gujarati* Chakalan; *Hindi* Andhali Chakalan, Do-mu Samp; *Kannada* Italiqi Havu; *Malayalam* Iruthalamoori; *Marathi* Do-tondya; *Oriya* Do-mundia Bora; *Tamil* Iruthalai Manniyan, Mannulli; *Telugu* Nalla Poda Pamu; *Urdu* Du-muhi)

DESCRIPTION Body short, stout, cylindrical; dorsals smooth; head indistinct from neck; snout broad, depressed; chin with mental groove; nostrils slit-like; eyes small; head scales small, only slightly larger than dorsals, slightly keeled; spur on each side of vent; tail short, blunt; dorsum of adults grey, brownish-red or dark brown, uniformly coloured or with indistinct black spots, especially towards and on tail; belly paler; juveniles have series of dark spots or cross-bars. **DISTRIBUTION** Western and peninsular India, Nepal and Pakistan; also Afghanistan and Iran. **HABITAT AND HABITS** Found in dry countryside such as deserts and coastal scrub, and associated with loose, sandy soil. Crepuscular and nocturnal. Diet includes small mammals and birds. Ovoviviparous; clutches comprise 6–8 young.

Tartary Sand Boa ▪ *Eryx tataricus* 34cm
(*Urdu* Tatar du-muhi)

DESCRIPTION Body short, stout, cylindrical; dorsals smooth anteriorly, slightly keeled posteriorly; head indistinct from neck; snout rounded; nostrils and eyes small; pair of spurs on each side of vent; tail short, tapering to acute point; dorsum yellowish-brown, paler on flanks. **DISTRIBUTION** Balochistan, western Pakistan. **HABITAT AND HABITS** Found in dry regions with sandy terrain and sparse grass and bushes. Active at dusk and at night. Diet includes lizards, birds and small mammals. Reproductive habits unknown.

◾ Boas/Typical Snakes ◾

Whitaker's Sand Boa ◾ *Eryx whitakeri* 79cm
(*Kannada* Irr-thale)

DESCRIPTION Body stout; head small, distinct from body; forehead scales small, smooth; mental groove absent; males have claw-like spur; tail short with blunt tip; dorsum brownish-grey or dark brown, with darker blotches that are fused and extend up to midbody, forming bands; belly unpatterned grey. **DISTRIBUTION** South-western India. **HABITAT AND HABITS** Found in scrub forests and on sea beaches. Fossorial. Diet includes rodents, snakes and birds. Reproductive habits unstudied.

COLUBRIDAE (Typical Snakes)
Until recently, this family was the largest of all the snake families, containing nearly 70 per cent of global snake species. Recent molecular work has helped reclassify the world's snake fauna, creating a number of families and reducing the number of species within the Colubridae. Nonetheless, this continues to be a large family (currently 1,861 species), with a great diversity of morphology and accompanying ecological characteristics, from reproductive modes and diets to the diversity of habitats occupied. Representatives include subfossorial, terrestrial and arboreal forms. All are non-venomous, though a few, such as cat snakes and kukri snakes, possess enlarged teeth and can inflict painful bites.

■ TYPICAL SNAKES ■

Common Vine Snake ■ *Ahaetulla nasuta* 200cm
(*Assamese* Beji Mukhua, Pat Hap; *Bengali* Laudoga; *Gujarati* Lile-jad Samp; *Hindi* Hara Samp; *Kannada* Hasiru Havu; *Malayalam* Pacha Pambu; *Marathi* Harantoli; *Oriya* Laudankia; *Punjabi* Hara Sapp; *Sinhalese* Ahaetulla; *Tamil* Kankuththi Pambu, Pachchai Pambu; *Telugu* Pasirika Pamu)

DESCRIPTION Body long, slender; snout pointed with dermal appendage; distinct middorsal groove on snout; pupils form horizontal slits; dorsum usually bright green, with longitudinal yellowish line along outer margin of ventrals. **DISTRIBUTION** India, Bangladesh, Bhutan, Nepal and Sri Lanka; also Southeast Asia. **HABITAT AND HABITS** Found in lightly forested areas, including gardens, in lowlands as well as mid-hills. Diet includes tadpoles, lizards, birds, small mammals and even leeches. Ovoviviparous, producing 3–23 young, after 172 days.

Oriental Vine Snake ■ *Ahaetulla prasina* 185cm
(*Assamese* Kacha Hap, Pat Hap, Seujia Hap; *Manipuri* Naril)

DESCRIPTION Body long, slender; snout not as long as Common Vine Snake's (see above); eyes have horizontal pupils; distinct groove in front of eyes; dorsum usually green, sometimes brown and yellow; yellow stripe along flanks of body; belly light green. **DISTRIBUTION** North-eastern India, Bangladesh and Bhutan; also Southeast Asia. **HABITAT AND HABITS** Found in forests, and associated with shrubs and saplings; also enters parks and gardens in search of lizards and birds. Ovoviviparous, producing 4–6 young.

■ Typical Snakes ■

Dice-like Trinket Snake ■ *Archelape bella* 80cm

DESCRIPTION Body slender; head indistinct from neck; snout rounded, with large rostral; eyes small; dorsals smooth; tail short and prehensile; dorsum red with greyish-brown to brown, saddle-shaped blotches, edged with black, covering 3–4 scales; saddle-like pattern less distinct on flanks; forehead has black 'Y'-shaped pattern. **DISTRIBUTION** North-eastern India; also Southeast Asia and southern China. **HABITAT AND HABITS** Found in submontane forests at 1,500–2,200m. Diet and reproductive habits unknown.

Iridescent Snake ■ *Blythia reticulata* 41cm
(*Miri* Dongkal-tabuvi)

DESCRIPTION Body stout; head indistinct from body; dorsals smooth; tail short and pointed; loreal and preocular scales absent; dorsum olive to dark, highly iridescent; dorsal scales sometimes light specked or bordered; juvenile dorsum has cream border with gap on vertebral line. **DISTRIBUTION** North-eastern India and Bangladesh; also adjacent Myanmar. **HABITAT AND HABITS** Found in evergreen forests at low to medium altitudes. Semifossorial, and associated with leaf litter. Oviparous.

■ Typical Snakes ■

Sri Lankan Cat Snake ■ *Boiga ceylonensis* 132cm
(*Sinhalese* Nidi Mapila; *Tamil* Komberi Muken)

DESCRIPTION Body slender, compressed; head distinct from neck; vertebral scales strongly enlarged; eyes large, pupils vertical; dorsum light tan; transverse bar on neck meets long streak on nape; series of alternating dark brown and pale chevrons; belly cream flecked with brown. **DISTRIBUTION** Western Ghats of south-western India, and Sri Lanka. **HABITAT AND HABITS** Found in evergreen forests, and associated with shrubs as well as the forest floor. Nocturnal and arboreal. Diet consists of geckos and agamid lizards. Clutches comprise 3–10 eggs, 25–29 x 8–13mm. Hatchlings 287–326mm.

Green Cat Snake ■ *Boiga cyanea* 187cm
(*Bengali* Sabuj Phanimonosha)

DESCRIPTION Body slender, compressed; head large, distinct from neck; eyes large, pupils vertical; centre of back has low ridge; vertebral scale row has enlarged scales; dorsals smooth; dorsum green; interstitial skin black; lower lips and chin pale blue; belly greenish-white or greenish-yellow, unpatterned or spotted with dark green; juveniles bright reddish-brown or olive, with green forehead and black postocular streak. **DISTRIBUTION** Eastern and north-eastern India, Nicobar Islands, Bangladesh, Bhutan and Nepal; also Southeast Asia. **HABITAT AND HABITS** Found in forest edges and disturbed habitats. Nocturnal and arboreal. Diet consists of frogs, birds and their eggs, lizards, snakes and small mammals. Clutches comprise 4–10 eggs, 40–48 x 15–21mm. Incubation period 64–83 days.

■ Typical Snakes ■

Forsten's Cat Snake ■ *Boiga forsteni* 231cm
(*Bengali* Brindabani Banka; *Oriya* Dalua Naga; *Sinhalese* Kabara Mapila, Maha Napila; *Tamil* Perunjoovai, Poonai Pambu)

DESCRIPTION Body slender, compressed; head distinct from neck and more or less triangular; eyes large, pupils vertical; dorsals smooth; vertebral scale row enlarged; dorsum greyish-brown; series of large brown cross-bars anteriorly; vertebral scales enlarged; dorsally greyish-brown, speckled or marbled with darker brown; sometimes reddish-brown, with or without dark cross-bars with pale spots in between; black stripe from frontal to nape, and sometimes 2 more stripes on nape; belly uniform cream or spotted with brown.

DISTRIBUTION Peninsular India, Nepal and Sri Lanka. **HABITAT AND HABITS** Found in forests, as well as agricultural fields in lowlands. Nocturnal and arboreal. Diet includes lizards, snakes, birds, bats and rodents. Clutches comprise 5–10 eggs.

Eastern Cat Snake ■ *Boiga gokool* 120cm

DESCRIPTION Body slender, compressed; head triangular, distinct from neck; eyes large, pupils vertical; dorsals smooth; dorsum yellowish-brown; usually, paired series of 'T'- or 'Y'-shaped marks on flanks; forehead has large, arrowhead-shaped brown mark, edged with black; black subocular stripe to angle of jaws; labials brown; belly cream with near continuous series of longitudinal stripes, composed of spots on each side of ventral. **DISTRIBUTION** Eastern and north-eastern India, Bangladesh and Bhutan. **HABITAT AND HABITS** Found in alluvial grassland, forests and plantations. Nocturnal and arboreal. Associated with bushes and shrubs. Diet includes rodents. Clutches comprise 8 eggs.

■ Typical Snakes ■

Many-spotted Cat Snake ■ *Boiga multomaculata* 99cm
(*Bengali* Phanimonosha; *Manipuri* Narilarangba)

DESCRIPTION Body slender, compressed; dorsals smooth; eyes large, pupils vertical; dorsum grey-brown, with black line from back of eye to jaws, and series of rounded, dark brown blotches; belly greyish-brown, with small brown spots. **DISTRIBUTION** North-eastern India and Bangladesh; also eastern China and Southeast Asia. **HABITAT AND HABITS** Found in lowland and submontane forests, at up to 1,500m. Nocturnal and arboreal, inhabiting short trees, bushes and bamboo groves. Diet includes lizards. Clutches comprise 5–7 eggs. Incubation period 60 days.

Tawny Cat Snake ■ *Boiga ochracea* 110cm
(*Manipuri* Narilasungba)

DESCRIPTION Body slender, compressed; head large, distinct from neck; eyes large, pupils vertical; dorsals smooth; vertebral scale row enlarged; dorsum reddish-brown, ochre or coral-red; unpatterned or poorly defined dark transverse line may be present. **DISTRIBUTION** Eastern Himalayas and north-eastern India, Bangladesh, Bhutan and Nepal; also Myanmar and Thailand. **HABITAT AND HABITS** Found in forests, as well as parks and gardens. Nocturnal and/or crepuscular, and arboreal. Diet comprises birds and their eggs, small mammals and lizards. Oviparous.

▪ Typical Snakes ▪

Assamese Cat Snake ▪ *Boiga quincunciata* 158cm

DESCRIPTION Body slender, compressed; head distinct from neck; eyes large, pupils vertical; dorsals smooth; vertebrals enlarged; dorsum yellow or greyish-brown, finely speckled with dark brown; vertebral series of dark brown or black spots or blotches; scales edged with white; flanks speckled or spotted with brown; nape has 3 longitudinal stripes; forehead brown; frontal and parietal scales black, edged with white; black postocular stripe to angle of jaws; belly yellowish-white, speckled with brown. **DISTRIBUTION** North-east India; also Southeast Asia. **HABITAT AND HABITS** Found in evergreen forests, especially in bamboo internodes, and also in tea plantations. Nocturnal. Diet unstudied. Clutches comprise 5 eggs, 45 x 13mm.

Thai Cat Snake ▪ *Boiga siamensis* 170cm

DESCRIPTION Body slender, compressed; head distinct from neck; eyes large, pupils vertical; dorsals smooth; vertebrals distinctly enlarged; dorsum light brown with squarish dark bands; flanks have dark and light alternating spots; belly light brown; forehead unpatterned; dark streak from posterior margin of eye to beyond last supralabial; 2 black stripes on either side of vertebral row to first dark band and continuous with it; subcaudals dark mottled. **DISTRIBUTION** Eastern and north-eastern India, Bangladesh and Bhutan; also Southeast Asia. **HABITAT AND HABITS** Found in evergreen forests, at up to 1,780m. Nocturnal and arboreal. Diet comprises rodents, birds and birds' eggs. Reproductive habits unstudied.

▪ Typical Snakes ▪

Common Indian Cat Snake ▪ *Boiga trigonata* 125cm

(*Bengali* Bankaraj, Phoni Monosha; *Gujarati* Kodiyo Samp; *Hindi* Dum Daraz Haphai; *Malayalam* Churla, Puchakannan Pambu; *Manipuri* Narilangangba; *Marathi* Manjar Sarp; *Nepali* Biralee Sarpa; *Oriya* Dhalua Naga; *Punjabi* Kikkri Sapp; *Sindhi* Landria; *Sinhalese* Garandi Mapila, Ran Mapila; *Tamil* Poonai Pambu, Wollai Churuttai; *Telugu* Tar Tutta; *Urdu* Maidani Billi-chisham)

DESCRIPTION Body slender, compressed; head distinct from neck; eyes large, pupils vertical; dorsals smooth; vertebral scales feebly enlarged; dorsum yellow to greyish-brown, with light grey, black-edged, arrowhead-shaped markings that may form vertebral stripe; light grey, black-edged, 'Y'-shaped mark on forehead; narrow dark streak, bordered above with light grey, from behind eye to mouth; belly white or grey, speckled with dark grey or black spots.
DISTRIBUTION Peninsular India, Bangladesh, Nepal, Pakistan and Sri Lanka; range extends west to the Middle East. **HABITAT AND HABITS** Found in forests as well as parks and gardens, and enters houses. Nocturnal and arboreal. Diet includes insects, lizards and birds and their eggs, and small mammals. Clutches comprise 3–11 eggs, 30–39 x 10–15mm. Incubation period 35–43 days. Hatchlings 24–26cm.

Nicobarese Cat Snake ▪ *Boiga wallachi* 105cm

(*South Nicobarese* Bo-aouna)

DESCRIPTION Body slender, compressed; head small, distinct from neck; snout long, greater than eye and projecting beyond lower jaw; eyes large, pupils vertical; dorsals smooth; dorsum yellowish-brown, each scale with brownish-olive tinge; lips yellow; no dark stripe behind eye; belly yellow, with dark blotches on abdomen and under tail.
DISTRIBUTION Great Nicobar, India.
HABITAT AND HABITS Found in lowland rainforests. Nocturnal and terrestrial. Diet includes birds' eggs. Reproductive habits unknown.

◾ TYPICAL SNAKES ◾

Ornate Flying Snake ◾ *Chrysopelea ornata* 175cm
(*Assamese* Maniraj, Nagraj; *Bengali* Bidjhori, Kalindi, Kalnagini, Uranta Shap; *Gujarati* Kali Nagin; *Malayalam* Parakunna Pambu; *Marathi* Udta Sonsarpa; *Sinhalese* Mal Karawala, Mal Sara, Polmal Karawala; *Tamil* Parakkum Pambu; *Urdu* Kala Jin)

DESCRIPTION Body slender; head depressed; eyes large, pupils rounded; ventrals with pronounced keels laterally; vertebral scales not enlarged; dorsals smooth or feebly

keeled; dorsum greenish-yellow or pale green; orange or red spots between dark cross-bands; head black dorsally, with yellow-and-black cross-bars; belly pale green with series of black lateral spots on each side. **DISTRIBUTION** Northern to north-eastern India, and southern Gujarat and Western Ghats in India, Bangladesh and Nepal; also Southeast Asia. **HABITAT AND HABITS** Found in old-growth trees within secondary vegetation and cultivated areas; may enter human habitation. Diurnal and arboreal. Known to make long (50m) glides from one tree to another, with body flattened. Diet includes geckos, agamid lizards, snakes, bats, rodents and birds. Clutches comprise 6–20 eggs, 26–38 x 13–18mm. Incubation period 65–80 days. Hatchlings 150–260mm.

Sri Lankan Flying Snake ◾ *Chrysopelea taprobanica* 100cm
(*Sinhalese* Dangara Danda)

DESCRIPTION Body slender; head depressed; eyes large, pupils rounded; ventrals have pronounced keels laterally; vertebral scales not enlarged; dorsals smooth or feebly keeled; dorsum greenish-yellow or pale green; no orange or red spots on midline; head black

dorsally, with irregular-edged yellow-and-black cross-bars on dorsum; belly pale green with black lateral spots on each side. **DISTRIBUTION** Eastern and Western Ghats, south-western India and northern Sri Lanka. **HABITAT AND HABITS** Found in old-growth trees, as well as secondary vegetation and cultivation, at up to 200m. Diurnal and arboreal. Diet comprises geckos, agamid lizards, bats, rodents, birds and snakes. Reproductive habits unknown.

■ TYPICAL SNAKES ■

Yellow-striped Trinket Snake ■ *Coelognathus flavolineatus* 180cm
(*Karen* Wukle)

DESCRIPTION Body relatively slender; snout long; eyes large, pupils rounded; dorsals keeled; tail about a fourth of snout-vent length; dorsum brownish-grey or olivaceous, with dark postocular stripe extending to above back of mouth, and another along nape; several short dark stripes or elongated blotches on top and sides of body. **DISTRIBUTION** Andaman Islands, India and Sunderbans of India and Bangladesh; also Southeast Asia. **HABITAT AND HABITS** Found in forests as well as open areas, such as parks and gardens. Diurnal and terrestrial. Diet includes rodents, birds, frogs and lizards. Clutches comprise 5–12 eggs. Incubation period 75–90 days.

Indian Trinket Snake ■ *Coelognathus helena* 168cm
(*Bengali* Arbaki Shap, Kumro-doga; *Gujarati* Rupsundri; *Malayalam* Kaatu Pambu; *Marathi* Taskar; *Oriya* Donger Chiti; *Sinhalese* Kalu Kateya; *Tamil* Kattu Pambu, Kattu Viriyan; *Telugu* Megarukula Poda)

DESCRIPTION Body slender; snout long; eyes large; dorsals smooth on flanks, weakly keeled posteriorly; dorsum brown or olive, with transverse bands and/or blotches on sides; neck has 2 narrow dark lines or dark-edged white collar; 2 dark stripes run along sides of body and tail. **DISTRIBUTION** India, Bangladesh, Nepal and Pakistan. **HABITAT AND HABITS** Found in forested plains and hills at up to 900m. Diurnal and terrestrial. Diet includes rats, lizards and birds. Clutches comprise 3–12 eggs.

▪ Typical Snakes ▪

Copper-headed Trinket Snake ▪ *Coelognathus radiatus* 230cm
(*Assamese* Bankaraj, Dhunduli Pheti, Goom Pheti; *Bengali* Dudhraj; *Manipuri* Tanglei Wachetmanbi)

DESCRIPTION Body relatively slender; snout long; dorsals keeled; eyes large, pupils rounded; dorsum greyish-brown or yellowish-brown, with 4 black stripes along front of body; cream stripe runs along upper 2 wide stripes; lower stripes narrower and may be broken up; head coppery-brown, with 3 radiating black lines from eyes. **DISTRIBUTION** India, Bangladesh, Bhutan and Nepal; also southern China and Southeast Asia. **HABITAT AND HABITS** Inhabits open areas such as grassland, ascending hills up to 1,400m. Diurnal and terrestrial. Diet includes birds and rats. Clutches comprise 5–12 eggs. Incubation period 70–95 days.

Defensive posture

Andaman Bronzeback Tree Snake
▪ *Dendrelaphis andamanensis* 130cm

DESCRIPTION Body slender; snout relatively broad, squarish; eyes large, pupils rounded; dorsum green-blue, with dark brown bands on anterior part of body; forehead olive; ventrolateral stripe along flanks absent; narrow postocular stripe; belly yellow. **DISTRIBUTION** Andaman Islands, India. **HABITAT AND HABITS** Found in lowland rainforests and enters human habitation. Diurnal and arboreal. Diet and reproductive habits unknown.

TYPICAL SNAKES

Stripe-tailed Bronzeback Tree Snake
■ *Dendrelaphis caudolineolatus* 915mm
(*Sinhalese* Viri Haldanda; *Tamil* Komberi Moorkhan)

DESCRIPTION Body slender; head distinct from neck; eyes large, pupils rounded; vertebrals weakly enlarged; forehead pale green; dorsum bronze-olive; front of body has oblique black streaks; narrow black temporal stripe; belly pale green or grey. **DISTRIBUTION** Sri Lanka and southern Western Ghats of India. **HABITAT AND HABITS** Found in relatively open secondary forests and home gardens. Diurnal and arboreal. Diet includes insects and frogs. Clutches comprise 3–6 eggs, 10 x 28–41mm.

Blue Bronzeback Tree Snake ■ *Dendrelaphis cyanochloris* 133cm
(*Assamese* Kar Hala; *Bengali* Lau Lata, Neel Shap)

DESCRIPTION Body slender; head distinct from body, slightly flattened; eyes large, pupils rounded; dorsals smooth; middorsals enlarged; ventrals distinctly keeled on sides; dorsum olive or bronze; scales edged with black; broad black stripe from head to beyond neck, breaking up into spots; ventrolateral stripe along flanks absent; belly pale green; interscale area of neck blue. **DISTRIBUTION** Eastern and north-eastern India, Andaman Islands, Bangladesh and Bhutan; also Southeast Asia. **HABITAT AND HABITS** Inhabits primary forests in lowlands and low hills, and enters agricultural areas. Diet includes lizards and possibly frogs. Reproductive biology unknown.

TYPICAL SNAKES

Nicobar Bronzeback Tree Snake ■ *Dendrelaphis humayuni* 620cm
(*South Nicobarese* Pichan)

DESCRIPTION Body slender; snout relatively broad, squarish; eyes large, pupils rounded; dorsum olive-brown, with scattered small black blotches; eyes white edged; dark lateral stripes along eyes and flanks; belly yellowish-olive, darkening at back; tail has black line on each side, and another on lower surface. **DISTRIBUTION** Nicobar Islands, India. **HABITAT AND HABITS** Found in lowland rainforests as well as more open grassland and within plantations. Diurnal and arboreal. Diet and reproductive habits unknown.

Painted Bronzeback Tree Snake ■ *Dendrelaphis proarchos* 121cm
(*Assamese* Kar Hala; *Bengali* Dora Gecho Shap, Lal-matha Talshara)

DESCRIPTION Body slender; head distinct from neck; eyes large, pupils rounded; vertebrals enlarged; tail nearly one-third total length; dorsals smooth and narrow except for 2 outer rows; anal single or divided; dorsum bronze-brown or brownish-olive, with black temporal stripe to neck; forehead brown with pale ventrolateral stripe, bordered with black stripes; blue or greenish-blue interscale area of neck. **DISTRIBUTION** Eastern and north-eastern India, Bangladesh and Nepal; also Southeast Asia and southern China. **HABITAT AND HABITS** Found in forested areas, also entering human habitation. Diurnal and arboreal. Diet includes frogs and geckos. Clutches comprise 5–8 eggs, 38 x 9mm.

Juvenile, close-up

Adult

▪ Typical Snakes ▪

Common Bronzeback Tree Snake ▪ *Dendrelaphis tristis* 150cm

(*Bengali* Lawjangli Shap, Shada-matha Talshara; *Gujarati* Oodto, Reliyo; *Hindi* Jard-ka-dhaman; *Malayalam* Villooni; *Marathi* Maniar, Rooka/Rookai; *Oriya*: Kauchia; *Sinhalese* Tura Haldanda; *Tamil* Chitooriki Pambu, Komberi Moorkan, Maram-eri Pambu, Padaiyeri Pambu, Panaiyeri Pambu; *Telugu* Chetturiki Pamu)

DESCRIPTION Body slender; head distinct from neck; eyes large, pupils rounded; tail long, third of total length; dorsals smooth; dorsum purplish- or bronze-brown; vertebral scales on neck and forebody yellow; buff flank-stripe from neck to vent; light blue on neck between scales revealed during display; belly pale grey, green or yellow. **DISTRIBUTION** India, Bangladesh, Nepal, Pakistan and Sri Lanka. **HABITAT AND HABITS** Found in secondary forests and forest clearings, and enters human habitation. Diurnal and arboreal. Makes long (up to 25m) jumps between trees. Diet consists of frogs and lizards, birds' eggs and insects. Clutches comprise 5–7 eggs, 29–39 x 10–12mm. Incubation period 2–6 weeks, hatchlings 21.6–33mm.

Mandarin Trinket Snake ▪ *Euprepiophis mandarinus* 170cm

DESCRIPTION Body stout; head short, scarcely distinct from neck; eyes large; tail short and stout; dorsum grey to greyish-brown, dorsum and tail have large, rounded yellow blotches, edged with black and yellow; belly and undertail cream coloured, sometimes with large, squarish black blotches. **DISTRIBUTION** North-eastern India; also southern China and Southeast Asia. **HABITAT AND HABITS** Inhabits forests from low altitudes to montane regions. Associated with rocky substrate, as well as scrub and agricultural fields. Diet consists of mice and shrews. Clutches comprise 3–8 eggs. Incubation period 42–55 days, hatchlings 300mm.

TYPICAL SNAKES

Khasi Hills Trinket Snake ■ *Gonyosoma frenatum* 150cm

DESCRIPTION Body slender, compressed; snout long; eyes large, pupils rounded; ventral keels developed; long, prehensile tail; dorsum grass-green to olive; lips light green; black postocular stripe to angle of jaws; belly pale green to white. **DISTRIBUTION** Northeastern India; also Southeast Asia and eastern China. **HABITAT AND HABITS** Found in evergreen and broadleaved forests, and associated with shrubs and low trees. Diet includes lizards, rats and birds. Reproductive habits unknown.

Red-tailed Trinket Snake ■ *Gonyosoma oxycephalum* 240cm

DESCRIPTION Body thick-set in adults, more slender in juveniles; head elongated, coffin shaped, slightly wider than neck; dorsals feebly keeled or smooth; dorsum emerald green, with pale green throat; black stripe along sides of head, across eyes; belly yellow; tail russet-brown; juveniles olive-brown with narrow white bars towards back of body. **DISTRIBUTION** Andaman Islands, India; also Southeast Asia. **HABITAT AND HABITS** Found in lowland rainforests. Arboreal as well as terrestrial. Diet includes rats, squirrels and birds; recorded in the Andamans entering limestone caves in search of bats. Clutches comprise 5–12 eggs, 65mm in length. Incubation period 100–120 days.

TYPICAL SNAKES

Green Trinket Snake ■ *Gonyosoma prasinum* 120cm

DESCRIPTION Body slender; snout rounded; head slightly distinct from neck; ventral keels developed; tail long; dorsum green, sometimes with brown tail-tip; lips and belly pale green; indistinct dark postocular stripe; chin cream coloured. DISTRIBUTION North-eastern India, and possibly also Nepal; also southern China and Southeast Asia. HABITAT AND HABITS Found in forested hills at up to 2,560m. Arboreal and associated with cane brakes and bamboo clumps. Diet includes lizards, birds and mammals. Clutches comprise 5–8 eggs. Incubation period 58–60 days; hatchlings 230mm.

Stripe-necked Snake ■ *Liopeltis frenatus* 76cm

DESCRIPTION Body slender, cylindrical; head not depressed; snout not projecting; tail relatively long; dorsals smooth; dorsum olive; scales edged with black, sometimes white, forming longitudinal stripes on front half of body; broad black stripe from back of eye to neck; upper lip and belly cream. DISTRIBUTION North-east India; also Southeast Asia. HABITAT AND HABITS Found in subtropical and montane forests at 600–1,800m. Terrestrial. Diet and reproductive habits unknown.

TYPICAL SNAKES

Stoliczka's Stripe-necked Snake ■ *Liopeltis stoliczkae* 600 mm

DESCRIPTION Body slender and cylindrical; head depressed, distinct from neck; snout projecting, twice as long as eye; pupil rounded; tail long and slender; dorsals smooth; anal divided; top of body brown or greyish-brown, with broad black stripe on sides of head extending to anterior body and fading thereafter; grey stripe on outer margins of ventrals; belly pale grey. **DISTRIBUTION** North-eastern India. Also Myanmar, Laos and Cambodia. **HABITAT AND HABITS** Found in evergreen and deciduous forested mid-hills (up to 700m asl). Arboreal, on bamboo. Diet and breeding habits unknown.

Common Wolf Snake ■ *Lycodon aulicus* 80cm

(*Assamese* Maruli; *Bengali* Gor Bonni Shap, Ghor Chitti; *Gujarati* Kodiyo Samp; *Hindi* Kauriwala; *Kannada* Choorta, *Malayalam* Shankuvariyan, Vellivaryan Pambu; *Marathi* Kandya; *Nepali* Sikham Phyancha; *Oriya* Heta Shapa, Jatia Shapa; *Punjabi* Kaudia; *Sinhalese* Alu Polonga, Tel Karawala; *Tamil* Kattu Viriyan; *Urdu* Chitra fraakh-dahan)

DESCRIPTION Body slender; head distinctly flattened; snout projects beyond lower jaw; dorsals smooth; dorsum dark brown or greyish-brown, with 12–19 white or pale yellow cross-bars, sometimes speckled with brown, which expand laterally; bands disappear posteriorly; upper lip cream coloured; belly cream or yellowish-white. **DISTRIBUTION** India, Bangladesh, Nepal, Pakistan and Sri Lanka; also Myanmar. **HABITAT AND HABITS** Found in parks and within human habitation, occupying roofs and cracks on walls. Diet includes geckos, snakes and rodents. Clutches comprise 3–11 eggs, 25–32mm. Hatchlings 140–190mm.

■ Typical Snakes ■

Sri Lanka Wolf Snake ■ *Lycodon carinatus* 60cm
(*Sinhalese* Dhara Karawala, Dhara Radanakaya)

DESCRIPTION Body slender; head distinctly flattened; snout projects beyond lower jaw; dorsals keeled; dorsum black with 19 distinct white rings, which may be reduced or absent in adults; black bands extend to belly, but diffuse. **DISTRIBUTION** Sri Lanka. **HABITAT AND HABITS** Found in forests, from lowlands to up to 1,500m. Hides under fallen leaves, logs and rubble by day. Diet includes frogs, geckos, skinks and non-venomous snakes, as well as lizard eggs. Clutches comprise 4–7 eggs. Mimic of the venomous Sri Lankan Krait (see p. 134).

Banded Wolf Snake ■ *Lycodon fasciatus* 85cm

DESCRIPTION Body slender, subcylindrical; head flattened; 2 enlarged posterior maxillary teeth; eyes small, pupils vertical; tail long; dorsals weakly keeled, keels more pronounced at back; dorsum glossy black, with 22–48 irregular cross-bars on body and tail, or reticulate or spotted pattern; belly blotched. **DISTRIBUTION** Eastern Himalayas and north-eastern India and Bangladesh; also Southeast Asia and southern China. **HABITAT AND HABITS** Found in evergreen forests at about 914–2,300m. Diet includes snakes, geckos and skinks. Clutches comprise 4–14 eggs. Hatchlings 216mm.

▪ Typical Snakes ▪

Andamans Wolf Snake ▪ *Lycodon hypsirhinoides* 72cm
(*Bengali* Ghar Chitti, Ghor-koney Shap; *Karen* Wu-sodi)

DISTRIBUTION Body slender; head flattened; snout rounded; scales smooth; dorsum dark brown, lacking banded pattern in adults; juveniles speckled dark brown and white;

no pale collar; belly cream or light yellow; upper lip scales without dark centres. **DISTRIBUTION** Andaman Islands, India. **HABITAT AND HABITS** Found in forested tracts, and enters human habitation. Diet includes lizards such as geckos and skinks. Clutches comprise 3–11 eggs. Incubation period 45 days.

Yellow-speckled Wolf Snake ▪ *Lycodon jara* 55cm
(*Bengali* Ghar-ginni Shap; *Manipuri* Naril)

DESCRIPTION Body slender; head flattened; ventrals not angular laterally; dorsum brown or purplish-black, finely stippled throughout with paired yellowish-white spots

or short longitudinal lines on each scale; upper lip and lower surface unpatterned white; white collar in juveniles. **DISTRIBUTION** Northern, eastern and north-eastern India, Bangladesh and Nepal. **HABITAT AND HABITS** Found in forested and agricultural areas. Nocturnal. Diet comprises frogs, lizards and small mammals. Oviparous.

▪ Typical Snakes ▪

Laotian Wolf Snake ▪ *Lycodon laoensis* 48cm

DESCRIPTION Body slender; forehead and lips deep blue; dorsum shiny black, with white-edged, yellow cross-bars, the first one on nape; cross-bars on body number 13–29, and 8–18 on tail, becoming narrower towards posterior; belly unpatterned cream. **DISTRIBUTION** North-eastern India; also southern China and Southeast Asia. **HABITAT AND HABITS** Found in evergreen forests in plains and low hills. Nocturnal. Diet includes frogs and lizards. Clutches comprise 5 eggs.

Large-toothed Wolf Snake ▪ *Lycodon septentrionalis* 180cm

DESCRIPTION Body slender, elongate, slightly compressed; head short, depressed; eyes moderate, pupils elliptical; 3 enlarged posterior maxillary teeth; tail long; dorsals smooth or median 4–7 rows weakly keeled; dorsum purplish-black, with 20–35 narrow white transverse bands on body and 10–17 on tail, which expand on flanks; belly white, sometimes spotted or barred with black; subcaudals with black speckling. **DISTRIBUTION** Eastern and north-eastern India and Bhutan; also Southeast Asia. **HABITAT AND HABITS** Found in low to mid-hills of evergreen forests at 220–2,100m, and often associated with streams. Diet comprises small vertebrates, including snakes. Reproductive habits unknown.

TYPICAL SNAKES

Barred Wolf Snake ■ *Lycodon striatus* 43cm
(*Oriya* Kaudia Chiti; *Sindhi* Abi-sangchul; *Sinhalese* Iri Karawala, Kabara Radanakaya; *Tamil* Utha Surita, Vellikkolvarayan; *Urdu* Chitra fraakh-dahan)

DESCRIPTION Body slender; snout obtusely rounded, somewhat flattened; head weakly differentiated from neck; supralabials 8, first and second of which contact nasal; dorsals smooth; forehead and dorsum black to dark brown, with series of white or yellow transverse marks; tail dorsum with irregular light longitudinal streaks; upper lips and belly unpatterned white. **DISTRIBUTION** Peninsular India, Nepal, Pakistan and Sri Lanka; range extends west to Central Asia. **HABITAT AND HABITS** Inhabits dry regions, including forest edges and semi-deserts, hiding under stones. Diet includes geckos and skinks. Clutches comprise 2–8 eggs, 25–30 x 9–12mm. Incubation period about 30 days.

Zaw's Wolf Snake ■ *Lycodon zawi* 70cm

DESCRIPTION Body slender; head flattened, distinct from neck; snout projecting; 2 enlarged posterior maxillary teeth; eyes small, pupils vertical; tail long; dorsals smooth; dorsum brownish-black, with narrow white bands that disappear posteriorly; lips pale brown; neck unbanded; belly cream coloured, each scale dark edged. **DISTRIBUTION** North-eastern India and Bangladesh; also adjacent Myanmar. **HABITAT AND HABITS** Found in forested lowlands and mid-hills, especially near streams, at up to 500m. Nocturnal and terrestrial. Diet includes skinks. Reproductive habits unstudied.

▪ Typical Snakes ▪

Afghan Awl-headed Snake ▪ *Lytorhynchus ridgewayi* 51cm
(*Afghan* Crotia-Sar Saamp)

DESCRIPTION Body slender; snout long and sharp; upper lips fail to contact eye; dorsum pale grey or pale brown, with median series of 40–49 blotches, dark brown or black anteriorly, turning pale brown with dark edges posteriorly; flanks have paired series of pale brown spots; forehead has anchor-shaped dark mark, contacting eye and corner of mouth; belly cream coloured. **DISTRIBUTION** Western Balochistan, Pakistan; range extends west to Afghanistan and Transcaspia. **HABITAT AND HABITS** Found in areas with gravel and scrub, and associated with roots of vegetation, at up to 2,000m. Diet includes arthropods and lizards. Clutches comprise 2–4 eggs.

White-barred Kukri Snake ▪ *Oligodon albocinctus* 76cm
(*Assamese* Patdei-hee; *Manipuri* Linjhak)

DESCRIPTION Body stout, cylindrical; head short, snout-tip blunt and rounded; eyes have rounded pupils; dorsals smooth; dorsum brownish-red, sometimes with white, yellow or fawn, black-edged cross-bars, numbering 19–27 on body and 4–8 on tail, or with dark brown, black-edged spots; forehead has dark stripe from upper lips to eyes; dark 'V'-shaped stripe from angle of mouth across parietals; belly cream, yellow or coral-red, marked with black. **DISTRIBUTION** Eastern Himalayas and north-eastern India, Bangladesh, Bhutan and Nepal; also Myanmar. **HABITAT AND HABITS** Found in forested habitats, and has also been seen in tea gardens. Crepuscular and terrestrial. Diet comprises insects, rats, frogs, and lizards and their eggs. Oviparous.

= TYPICAL SNAKES =

Banded Kukri Snake ■ *Oligodon arnensis* 64cm
(*Bengali* Udaykal; *Gujarati* Sankh Bangani; *Hindi* Kukri Samp; *Malayalam* Churute; *Manipuri* Linkhak; *Marathi* Gargar; *Nepali* Gurbay; *Oriya* Matia Hara Sapa; *Sinhalese* Arani Dath Ketiya; *Tamil* Kattu Viriyan, Olai Pambu, Pul Viriyan; *Telugu* Sanka; *Urdu* Patta Kukri Saamp)

DESCRIPTION Body stout, cylindrical; snout short and blunt; dorsals smooth; dorsum brown, usually with red or purple markings, lighter on flanks, with 32–41 black cross-bars

or spots that break up on flanks into streaks, sometimes edged with cream; bars 1–5 scales wide; head has 3 arrow-shaped dark marks; belly cream with indistinct lateral spots. **DISTRIBUTION** Peninsular India, Bangladesh, Nepal, Pakistan and Sri Lanka. **HABITAT AND HABITS** Found in forests, as well as parks and gardens. Diet includes rats, lizards and reptile eggs; also known to scavenge. Clutches comprise 3–9 eggs, 36 x 10mm.

Grey Kukri Snake ■ *Oligodon cinereus* 73cm

DESCRIPTION Body stout, subcylindrical; head short, indistinct from neck; eyes moderate, pupils rounded; tail short, blunt; dorsals smooth; dorsum reddish-brown or red, unpatterned, or with white or grey, black-edged cross-bars; forehead unpatterned brown; belly cream coloured, often with squarish dark spots. **DISTRIBUTION** North-east India and Bangladesh;

also Southeast Asia. **HABITAT AND HABITS** Inhabits lowlands and mid-hills at 120–890m. Nocturnal and terrestrial. Diet includes spiders and insects. Clutches comprise 4–5 eggs.

■ Typical Snakes ■

Cantor's Kukri Snake ■ *Oligodon cyclurus* 94cm
(*Bengali* Tukkr-bora)

DESCRIPTION Body stout, cylindrical; ventrals angular laterally; dorsals smooth; dorsum yellowish-brown or dark brown, with dark reticulation and black-edged cross-bars or transverse dark oval spots; head has dark, 'V'-shaped markings; belly cream coloured, and sometimes dark spotted. **DISTRIBUTION** Eastern and north-eastern India, Bangladesh and Nepal; also Southeast Asia. **HABITAT AND HABITS** Found in forests as well as agricultural areas. Diet includes snakes, lizards and their eggs. Clutches comprise 3–26 eggs, 23–31 x 16–19mm. Hatchlings 165–200mm.

Spot-tailed Kukri Snake ■ *Oligodon dorsalis* 50cm
(*Manipuri* Naril)

DESCRIPTION Body cylindrical and stout; head short, blunt; dorsals smooth; dorsum generally dark brown or purple, with light vertebral stripe, sometimes dark edged, or containing small black spots; second stripe along second and third dorsal scale rows; top of tail has 2–3 large black spots; forehead dark brown with 2 cross-bars; belly orange with squarish black spots. **DISTRIBUTION** North-eastern India, Bangladesh and Bhutan; also Southeast Asia. **HABITAT AND HABITS** Found in low hills in evergreen forests, at elevations from lowlands up to 2,675m. Diurnal. Diet and reproductive habits unknown.

▪ Typical Snakes ▪

Duméril's Kukri Snake ▪ *Oligodon sublineatus* 35cm
(*Sinhalese* Pulli Dath Ketiya)

DESCRIPTION Body cylindrical and stout; head short, blunt; dorsals smooth; dorsum pinkish-brown with small dark brown spots; belly light pink and brown, with 3 rows of

brown markings; 2 lateral rows of linear marks confluent and form stripes; median row of discontinuous spots ending at vent. **DISTRIBUTION** Sri Lanka. **HABITAT AND HABITS** Found in lightly forested areas, and enters human habitation, at up to 1,200m. Diet includes reptile eggs, small lizards and frog eggs. Clutches comprise 2–3 eggs.

Streaked Kukri Snake ▪ *Oligodon taeniolatus* 59cm
(*Bengali* Uday Kaal; *Sinhalese* Wairi Dath Ketiya; *Telugu* Wanapa Pam; *Urdu* Dahari-dar Kukri Saamp)

DESCRIPTION Body cylindrical and stout; head short, blunt; dorsals smooth; dorsum light brown, with 36–47 narrow black transverse cross-bars that may form irregular

spots, with or without 4 dark brown longitudinal stripes; sometimes, cream vertebral stripe; belly cream coloured. **DISTRIBUTION** Northern and peninsular India, Bangladesh, Pakistan and Sri Lanka. **HABITAT AND HABITS** Found in thinly forested areas, at elevations from sea level to about 200m. Nocturnal and terrestrial. Diet includes reptile and frog eggs, and lizards. Clutches comprise 3–9 eggs.

▪ Typical Snakes ▪

Black-banded Trinket Snake ▪ *Oreocryptophis porphyraceus* 125 cm

DESCRIPTION Body slender; head elongated, indistinct from neck; snout rounded; ventral keels absent; dorsum reddish-brown, body and tail dark banded; dark stripe along posterior half of body, and another on forehead and behind each eye; belly unpatterned cream or yellow. **DISTRIBUTION** Eastern Himalayas and north-eastern India, Bangladesh and Bhutan; also eastern China and Southeast Asia. **HABITAT AND HABITS** Found in montane forests, and in low vegetation of shrubs. Diet includes voles and shrews. Clutches comprise 2–5 eggs. Incubation period 50–60 days.

Eastern Trinket Snake ▪ *Orthriophis cantoris* 196cm

DESCRIPTION Body slender; head slightly distinct from neck; ventral keel distinct; dorsum olive with horizontal black bars separated by bands of white-edged scales; forehead brown; iris red; dorsal surface of tail has red spots; belly cream anteriorly, becoming reddish with darker mottling towards tail. **DISTRIBUTION** Eastern Himalayas and north-eastern India, Bhutan and Nepal; also Myanmar. **HABITAT AND HABITS** Found in montane regions at altitudes of 1,000–2,300m, where it inhabits forested habitats. Crepuscular. Diet likely to be small mammals and birds. Clutches comprise 10 eggs.

▪ Typical Snakes ▪

Himalayan Trinket Snake ▪ *Orthriophis hodgsonii* 210cm
(*Nepali* Karait, Pila Matia)

DESCRIPTION Body slender; snout rounded; ventral keels developed; dorsum olive-brown; some scales bordered with black and white, resulting in reticulate pattern;

large dark blotch on mid-forehead; belly yellow with dark spots. **DISTRIBUTION** Himalayas of northern and eastern India and Nepal; also southern China. **HABITAT AND HABITS** Found in moist forests at 1,000–3,200m, in addition to edges of agricultural fields in vicinity of water. Diet includes rats, toads and skinks. Oviparous.

Striped Trinket Snake ▪ *Orthriophis taeniurus* 270cm

DESCRIPTION Body slender, compressed; snout long; dorsals smooth; dorsum pale yellow or grey, with dark stripe on sides of head; body has scattered black blotches, and sometimes cream or yellow stripe down middle of back; forehead brown; posterior half of body has dark lateral stripe, interspaced with faint pale bands descending towards ventrals. **DISTRIBUTION** Eastern Himalayas and Andaman Islands, India and Bhutan; also eastern China and Southeast Asia. **HABITAT AND HABITS** Found in forests and fields, and often in limestone caves. Diet includes bats and rodents, as well as swiftlets and their eggs. Clutches comprise 6–15 eggs.

■ Typical Snakes ■

Mock Viper ■ *Psammodynastes pulverulentus* 55cm
(*Adi* Biya-biya; *Bengali* Pahari Shap; *Nepali* Gurbay)

DESCRIPTION Body slender; head flattened, distinct from neck; snout short, truncated in profile; eyes large, pupils vertical; dorsals smooth; dorsum reddish-brown to yellowish-grey, to black, with small dark spots or streaks; longitudinal stripe along middorsal region, and 3 longitudinal stripes along flanks; belly spotted with brown or grey, and with dark spots or longitudinal lines.

DISTRIBUTION Northern, eastern and north-eastern India, Bangladesh, Bhutan and Nepal; also Southeast Asia and southern China. **HABITAT AND HABITS** Found in low to middle-altitude forests and associated with streams. Teeth adapted for diet of heavily scaled prey such as skinks. Frogs, geckos and small snakes also consumed. Ovoviviparous, producing 3–10 young.

Eastern Rat Snake ■ *Ptyas korros* 220cm
(*Assamese* Rayel; *Manipuri* Tanglei)

DESCRIPTION Body slender; head elongate, distinct from neck; eyes large, pupils rounded; dorsals smooth; head and anterior of body dorsum grey to olive-brown, posterior of body darkening to nearly black; scales edged with white, appearing as white bands on black background; belly, chin and lips brownish-cream.
DISTRIBUTION North-eastern India, Bangladesh and Bhutan; also Southeast Asia and eastern China. **HABITAT AND HABITS** Inhabits lowland forests and also subtropical forests at up to 1,800m. Mostly terrestrial, and can also climb trees. Diet includes rodents. Clutches comprise 4–12 eggs. Incubation period 61 days. Hatchlings 290mm.

▪ Typical Snakes ▪

Indian Rat Snake ▪ *Ptyas mucosa* 370cm
(*Assamese* Holonga Gom, Hulaberia, Machoa Gom; *Bengali* Darash, Masua Alod; *Gujarati* Dhaman; *Hindi* Dhaman; *Kannada* Chere Havu, Kiyai Havu; *Malayalam* Chera, Sara; *Manipuri* Tanglei; *Marathi* Dhaman, Dhamin; *Nepali* Dhaman; *Oriya* Dhamana Shapa; *Punjabi* Matyala Chua Khana Sapp; *Sinhalese* Kalu Garandiya, Vola Garandiya; *Tamil* Sarai Pambu; *Telugu* Jerri Pothu; *Urdu* Dhaman, Dahaman)

DESCRIPTION Body slender; head elongate, distinct from neck; eyes large, pupils rounded; dorsals smooth; dorsum yellowish-brown, olivaceous-brown to black; posterior of

body has dark bands or reticulate pattern; lip and flank scales dark edged; belly greyish-white or yellow. **DISTRIBUTION** India, Bangladesh, Bhutan, Nepal, Pakistan and Sri Lanka; range extends from Turkmenistan and Iran, through the subcontinent, to southern China and Southeast Asia. **HABITAT AND HABITS** Found in forests, scrubland, agricultural fields, parks and cities. Diet includes frogs, rats, bats, birds, lizards, turtles and snakes. Clutches comprise 5–18 eggs, 50–57mm. Incubation period 60 days. Hatchlings 361–472mm.

Green Rat Snake ▪ *Ptyas nigromarginata* 184cm
(*Angami Naga* Paee; *Manipuri* Tanglei Asungba)

DESCRIPTION Body slender, slightly compressed; head elongate; eyes large; 4–6 median dorsals keeled; dorsum green or olive-green; dorsal scales edged with black; in juveniles, 4 longitudinal black stripes along body and tail, which in adults are on posterior third of body; head olive-brown, bright yellow temporal patch; belly with greenish-yellow tinge.

DISTRIBUTION North-eastern India, Bangladesh, Bhutan and possibly Nepal; also Southeast Asia and southern China. **HABITAT AND HABITS** Found in montane forests, as well as disturbed areas, at 1,485–2,300m. Terrestrial. Diet includes rats, birds, lizards and snakes. Clutches comprise up to 9 eggs.

▪ Typical Snakes ▪

Two-coloured Forest Snake ▪ *Rhabdops bicolor* 60cm

DESCRIPTION Body stout, cylindrical; head indistinct from neck, depressed; snout rounded; nostrils dorsolateral; eyes small; dorsals smooth; dorsum dark brown or black; belly yellowish-cream, the 2 colours forming line of demarcation; cream-coloured, unpatterned or white-black spots under tail. DISTRIBUTION North-eastern India; also adjacent southern China and Myanmar. HABITAT AND HABITS Found in forested mid-hills. Diet includes earthworms and slugs. Reproductive habits unknown.

Collared Black-headed Snake ▪ *Sibynophis collaris* 76cm

DESCRIPTION Body slender; head relatively short, slightly distinct from neck; pupils rounded; dorsals smooth; dorsally brown or greyish-brown, with black vertebral region comprising black spots; occasionally has light, dotted dorsolateral line; dark cross-bar behind eyes and on forehead; black transverse band on nape bordered posteriorly with yellow; belly yellow, each ventral with 2 outer black spots. DISTRIBUTION Northern, eastern and north-eastern India, Bangladesh, Bhutan and Nepal; also Southeast Asia. HABITAT AND HABITS Found in forests in hills and plains. Diet consists of skinks, snakes, frogs and insects. Clutches comprise up to 6 eggs.

▪ Typical Snakes ▪

Cantor's Black-headed Snake ▪ *Sibynophis sagittarius* 45cm
(*Bengali* Doblee, Itelchapa, Mathakalo Shap; *Urdu* Sunahra Saamp)

DESCRIPTION Body slender; head short, slightly distinct from neck; snout broad; pupils rounded; dorsals smooth; dorsally light brown, with dark brown flanks; vertebral series

of black dots; head and nape dark brown or black, with large, oval, elongate yellow mark on sides of back of head; lips, throat and belly yellow; black dot on outer edge of each ventral. DISTRIBUTION Northern and eastern India, Bangladesh, Bhutan, Nepal, Pakistan and Sri Lanka. HABITAT AND HABITS Found in forested areas and agricultural fields. Terrestrial. Diet includes frogs, blind snakes and lizards. Oviparous.

Black-headed Snake ▪ *Sibynophis subpunctatus* 42cm
(*Sinhalese* Dathigomaraya)

DESCRIPTION Body slender; head short, slightly distinct from neck; snout broad; pupils rounded; dorsals smooth; dorsum light brown with greyish-brown flanks; vertebral series

of black dots; head and nape dark brown or black, with large, oval, elongate patch of yellow at back of forehead; lips, throat and belly yellow; black spot on outer edge of ventral. DISTRIBUTION Peninsular India, Bangladesh and Sri Lanka. HABITAT AND HABITS Found in forested hills and agricultural fields at up to 750m. Terrestrial. Diet includes frogs, snakes and lizards. Clutches comprise 2–6 eggs.

▪ Typical Snakes/Sand Snakes ▪

Assamese Slender Snake ▪ *Trachischium monticola* 275mm

DESCRIPTION Body slender; head indistinct from neck; eyes small, pupils rounded; dorsals keeled on posterior of body and base of tail (males), or feebly keeled or smooth (females); 15 middorsal scale rows; dorsum light or dark brown, iridescent, with black longitudinal lines and 2 light or reddish-brown dorsolateral stripes; yellow spot on sides of neck; belly yellow. DISTRIBUTION North-eastern India, Bangladesh and Nepal; also Tibet. HABITAT AND HABITS Found in subtropical forests. Terrestrial. Diet and reproductive habits unknown.

LAMPROPHIIDAE (Sand Snakes)
The Lamprophiidae family includes 72 non-venomous snake species that were, until recently, placed in the Colubridae family. The body is cylindrical with smooth scales, and the eyes are large with rounded pupils. Sand snakes are terrestrial, and are predators of small vertebrates.

Pakistani Ribbon Snake ▪ *Psammophis leithii* 76cm
(*Gujarati* Pattawalo Samp)

DESCRIPTION Body slender; head elongated, narrow, distinct from neck; body cylindrical, rather slender; eyes large, pupils rounded; dorsum yellowish-brown, yellowish-grey or straw in colour, usually with 4 dark brown longitudinal stripes, the 2 median ones distinct and bordered with black spots that may extend to eyes; outer pair, when present, extends forwards to nostrils; belly unpatterned yellowish-cream. DISTRIBUTION Northern India and Pakistan. HABITAT AND HABITS Found in marshes, deserts and grassland, and associated with bushes, at below 800m. Diurnal, climbing bushes and low trees. Diet includes skinks and mice. Clutches comprise 4–10 eggs.

KEELBACK SNAKES

> **NATRICIDAE (KEELBACK SNAKES)**
> This family of aquatic or at least mesic-habitat snakes includes 231 described species. It contains one genus of venomous snake (*Rhabdophis*), associated with human mortality. All others are non-venomous. Some members show caudal autotomy, which is more familiar among lizards.

Buff-striped Keelback ■ *Amphiesma stolatum* 80cm
(*Assamese* Bamuni; *Bengali* Dora Shap, Hele Shap; *Gujarati* Pattawalo Samp; *Hindi* Hurva, Seeta-ki-latt; *Malayalam* Churrutay, Theyyan Pambu; *Manipuri* Lilha; *Marathi* Naneti; *Nepali* Harhare Sarpa; *Oriya* Bamhuni, Kandanala; *Sinhalese* Aharukukka, Magarakinna; *Tamil* Kaliyan Kutty, Nikitan Kutty, Nirkatan Pambu; *Telugu* Wanna Pamu; *Urdu* Lakeer-dar Khar-pusht)

DESCRIPTION Body moderately slender; head distinct from neck; dorsum olive-grey to greenish-grey; buff or orange stripes run dorsolaterally on fifth to seventh scale rows. **DISTRIBUTION** India, Bangladesh, Nepal, Pakistan and Sri Lanka; also Southeast Asia. **HABITAT AND HABITS** Found in grassland, as well as lowland forests, usually in vicinity of water such as lakes and ponds, as well as rice fields. Diet includes frogs, insects, insect larvae, scorpion, fish and lizards. Clutches comprise 3–15 eggs, 22–35 x 13–18mm. Incubation period 36–62 days.

Black-spined Snake ■ *Aspidura ceylonensis* 522mm
(*Sinhalese* Kurun Karawala, Rath Karawala)

DESCRIPTION Body slender and cylindrical; head long, snout broadly rounded; neck indistinct; tail short; scales smooth, iridescent; dorsal ground colour crimson-brown with black vertebral stripe; dorsum of forebody brown; laterally with series of black spots; neck region has distinct dark brown marking; belly crimson. **DISTRIBUTION** Central Highlands of Sri Lanka. **HABITAT AND HABITS** Found in submontane forests at around 1,300m, and sometimes enters plantations. Fossorial; associated with damp soil, silted-up drains and decaying leaves. Diet includes earthworms. Clutches comprise 2–5 eggs.

■ Keelback Snakes ■

Günther's Rough-sided Snake ■ *Aspidura guentheri* 160mm
(*Sinhalese* Kuda Madilla)

DESCRIPTION Body slender and cylindrical; head indistinct from neck; dorsum brown, mottled with dark brown; forehead dark; pale neck-band; vertebral and lateral rows of dark spots; belly light brown. DISTRIBUTION Sri Lanka. HABITAT AND HABITS. Known from lowland forests at 100–500m. Diet includes earthworms. Clutches comprise 1–3 eggs.

Common Rough-sided Snake ■ *Aspidura trachyprocta* 383mm
(*Sinhalese* Dalawa Madilla, Le Madilla)

DESCRIPTION Body slender and cylindrical; head indistinct from neck; dorsum ranges from blackish-brown to light reddish-brown, with lateral stripes and 2–3 rows of dark dorsal spots; belly colour ranges from heavily blotched with black, to light yellow with red tint and blotched on midline. DISTRIBUTION Sri Lanka. HABITAT AND HABITS Found in leaf litter, soil and rotten timber, in forests and agricultural land, at 750–2,100m. Diet unknown. Clutches comprise 4–12 eggs, 16 x 25mm.

■ Keelback Snakes ■

Olive Keelback Water Snake ■ *Atretium schistosum* 100cm
(*Bengali* Kerul Metuli, Mete Shap; *Hindi* Hara Pani Ka Samp; *Kannada* Barmnya; *Malayalam* Neerkoli; *Sinhalese* Diyawarna, Kadola; *Tamil* Pachchai Tanni Pambu; *Telugu* Nalla Wahlagilee Pam)

DESCRIPTION Body slender; snout short; nostrils slit-like, placed on top of snout; scales distinctly keeled, and strongest on posterior part of body and tail; dorsum olive-brown

or greenish-grey, unpatterned or with 2 series of small black spots; occasionally, dark lateral streak; some individuals have red lateral line; outer row of scales and belly yellow, cream or red. **DISTRIBUTION** India, Bangladesh, Nepal and Sri Lanka. **HABITAT AND HABITS** Found in lowlands associated with wetlands, at up to about 1,000m. Diet includes frogs, tadpoles, fish, crustaceans and crabs. Clutches comprise 10–32 eggs, 15 x 20–25mm. Hatchlings 167–174mm.

Beddome's Keelback ■ *Hebius beddomii* 69cm

DESCRIPTION Body slender; scales keeled; dorsum olive-brown, with series of yellow spots, enclosed by 2 black spots or short bars; oblique, black-edged yellow stripe from eye to angle

of jaws; lips yellow; belly cream coloured, unpatterned or dotted with brown on flanks; juveniles have yellow collar. **DISTRIBUTION** Western Ghats, south-western India. **HABITAT AND HABITS** Found in evergreen and moist deciduous forests, in mid-hills. Diurnal and crepuscular, and active at edges of water bodies. Diet includes toads. Reproductive habits unknown.

KEELBACK SNAKES

Clerk's Keelback ■ *Hebius clerki* 77cm

DESCRIPTION Body slender; scales strongly keeled; scales on first dorsal row keeled; tail long; dorsum dark brown or greyish-brown, becoming progressively darker on back; 2 pale dorsolateral stripes or series of spots on back and tail; forehead brown; short yellow vertebral streak behind occiput; postocular streak not interrupted at level of neck; reddish-brown spots on flanks; belly unpatterned yellow with row of black dots on each side. **DISTRIBUTION** Eastern and north-eastern India and Nepal; also Southeast Asia and southern China. **HABITAT AND HABITS** Found in forests in mid-hills. Diet unknown. Oviparous.

Khasi Hills Keelback ■ *Hebius khasiense* 60cm

DESCRIPTION Body slender; head distinct from neck; eyes large, pupils rounded; dorsals keeled; dorsum has 3 blackish-brown stripes, alternating with 4 reddish-brown stripes; forehead greyish-brown with pale brown marks; lips white with dark edges, continuing to sides of neck; upper lips at back of head have rounded blotches; belly white; ventrals have reddish-brown outer edges. **DISTRIBUTION** North-eastern India; also Southeast Asia and southern China. **HABITAT AND HABITS** Found in submontane forests at 900–1,000m. Associated with forest floor, near hill streams. Diet comprises insects, frogs and tadpoles. Oviparous.

■ Keelback Snakes ■

Boulenger's Keelback ■ *Hebius parallelum* 64cm
(*Nepali* Pani Samp)

DESCRIPTION Body slender; scales weakly keeled; scales on first dorsal row smooth; tail short; dorsum olive-brown or greyish-brown, with 2 light dorsolateral stripes or series of spots on back and tail; forehead brown; short yellow vertebral streak behind occiput; postocular streak interrupted at level of neck; belly unpatterned yellow with row of black dots on each side. **DISTRIBUTION** North-eastern India. **HABITAT AND HABITS** Inhabits forests in mid-hills. Diet unknown. Oviparous.

Venning's Keelback ■ *Hebius venningi* 68cm

DESCRIPTION Body relatively slender; scales feebly keeled; outer scale rows smooth; dorsum greyish-brown, with indistinct black squares; incomplete collar may be present; juveniles have chain-like yellow markings on flanks; belly cream coloured, edged with dark brown. **DISTRIBUTION** North-eastern India; also southern China and Southeast Asia. **HABITAT AND HABITS** Found in forested habitats in mid-hills. Diet includes tadpoles. Reproductive habits unknown.

▪ Keelback Snakes ▪

Strange-tailed Keelback ▪ *Hebius xenura* 66cm

DESCRIPTION Body slender; head distinct from neck; eyes large, pupils rounded; dorsals keeled; subcaudals 81–107, entire; dorsum olive-brown to nearly black; paired series of reddish-orange, pale brown, yellow or white spots on flanks, and adjacent spots sometimes connected by faint black cross-lines; lips white, with dark lines on sutures; belly white or yellow; outer edges of ventrals have dark brown spots; subcaudals dark grey. DISTRIBUTION Northeastern India and Bangladesh; also Myanmar. HABITAT AND HABITS Found in evergreen forests. Associated with forest streams. Diet and reproductive biology unknown.

Eastern Keelback ▪ *Herpetoreas platyceps* 88cm
(*Nepali* Matia; *Urdu* Chitra Khar-pusht)

DESCRIPTION Body slender; dorsum dark grey, bronze-brown, reddish-brown or yellowish-brown; sometimes has dark lateral stripe; dark band from rostral through eye to last supralabial; nuchal loop sometimes present; belly yellowish-cream or with orange tinge, sometimes with small dark streaks, especially posteriorly; throat often powdered with black. DISTRIBUTION India, Bangladesh, Nepal and Pakistan. HABITAT AND HABITS Found in forests and their edges, and enters agricultural fields and human dwellings. Diet consists of skinks and frogs. Clutches comprise 2 eggs, 8 x 25mm.

◾ Keelback Snakes ◾

Green Keelback ◾ *Macropisthodon plumbicolor* 485mm
(*Bengali* Sabuj Dhora; *Gujarati* Leelo Samp; *Hindi* Hara Nag; *Malayalam* Pacha Moorkham; *Marathi* Gautya Samp; *Sinhalese* Palabariya, Pala Polonga; *Tamil* Pachcha Neerkkoli)

DESCRIPTION Body stout; head distinct from neck; eyes moderately large, pupils rounded; tail short; dorsals have strong keels; dorsum bright green; juveniles have 'V'-shaped mark on nape; black postocular stripe to angle of jaws; transverse black spots or cross-bars on dorsum and tail; belly cream, yellow or grey. **DISTRIBUTION** India, Bangladesh and Sri Lanka. **HABITAT AND HABITS** Found in lowlands associated with wetlands, at up to 1,000m. Diet comprises frogs and toads. Eggs 21–36mm. Hatchlings 136–168mm.

Himalayan Keelback ◾ *Rhabdophis himalayanus* 125cm
(*Manipuri* Lilha)

DESCRIPTION Body stout; head distinct from neck; eyes large, pupils rounded; dorsals keeled; dorsum olive, olive-brown or dark brown, with 2 dorsolateral rows of orange-yellow spots, more numerous anteriorly; anterior body chequered; neck has yellow or orange collar, edged with black; black postocular stripe; lips yellow, edged with black; belly yellowish-white, darker towards tip. **DISTRIBUTION** Eastern and north-eastern India, Bangladesh and Nepal; also southern China and Southeast Asia. **HABITAT AND HABITS** Inhabits forests, particularly on rocky slopes, and enters agricultural fields and vicinity of streams. Diet comprises frogs, lizards and fish. A back-fanged snake with a bite that may have serious effects on humans. Oviparous.

■ Keelback Snakes ■

Collared Keelback ■ *Rhabdophis nuchalis* 62cm

DESCRIPTION Body stout, cylindrical; head flattened, distinct from neck; nuchal groove distinct; eyes large, pupils rounded; dorsals keeled, except smooth outer row; dorsum light brown, checkered with pale reddish-brown spots; forehead brown, speckled with red; 2 oblique black stripes between back of upper lips; alternate rows of body scales red and brown; fine, oblique black lines on dorsolateral scales at midbody; neck reddish-brown from angle of jaw; juvenile has reddish-yellow collar, a reddish tinge at back. **DISTRIBUTION** North-eastern India; also southern China. **HABITAT AND HABITS** Found in secondary subtropical broadleaved forests, as well as terrace cultivation, at 1,500–1,750m. Diet unknown. Clutches comprise 8–19 eggs, 26 x 13mm.

Red-necked Keelback ■ *Rhabdophis subminiatus* 130cm
(*Assamese* Halikajopia, Rongadingi Gom; *Bengali* Laldora; *Manipuri* Chingkharou, Tanglei Kokngangbi)

DESCRIPTION Body slender; head distinct from neck; eyes large, pupils rounded; dorsals strongly keeled; anal divided; dorsum olive-brown or green, unpatterned or with black-and-yellow reticulation; nape has yellow-and-red band; oblique dark bar below eye; belly yellow, sometimes with black dot on outer end of each shield; juveniles have black cross-bar or triangular mark on nape, bordered with yellow behind. **DISTRIBUTION** North-eastern India, Bangladesh and Nepal; also southern China and Southeast Asia. **HABITAT AND HABITS** Found in forests in hills areas as well as lowlands, in vicinity of ponds and streams. Diet consists of frogs and toads. A rear-fanged snake with a venom gland. Clutches comprise 5–17 eggs, 17.5–27 x 11–15mm. Incubation period 50 days. Hatchlings 130–190mm.

KEELBACK SNAKES

Sri Lankan Keelback Water Snake ■ *Xenochrophis asperrimus* 89cm
(*Sinhalese* Diya Bariya, Diya Polonga; *Tamil* Tanni Pambu)

DESCRIPTION Body stout; head distinct from neck; eyes small to medium, pupils rounded; nostrils directed slightly upwards; dorsal scales strongly keeled; dorsum olive-brown, with 22–32 black spots or cross-bars, which are distinct in anterior half of body; belly yellowish-green to yellowish-ochre. **DISTRIBUTION** Sri Lanka. **HABITAT AND HABITS** Found in wetlands in plains, including rice fields, ponds, lakes, marshes and rivers, at 1,000m. Diet includes fish and frogs. Clutches comprise 4–30 eggs, 15–16 x 27–30mm. Hatchlings 100mm.

Dark-bellied Keelback Water Snake ■ *Xenochrophis cerasogaster* 51cm
(*Assamese* Dhor, Dhora; *Bengali* Kalo Mete, Lal Mitallee; *Jogi* Meeka; *Urdu* Surakh Dhoobi Saamp)

DESCRIPTION Body slender; head narrow and blunt; scales strongly keeled; dorsum olive-brown, brown or green, sometimes with distinct dark spots or pale yellow stripes; lips yellow; belly reddish-grey, with brown or purplish-black blotches as well as white spots; outer margins of ventrals marked with yellow line, especially in juveniles; chin and throat white with red mottling. **DISTRIBUTION** Northern and north-eastern India, Bangladesh, Nepal and Pakistan. **HABITAT AND HABITS** Found at edges of marshes and swamps, in side pools with emergent vegetation. Diet includes fish and shrimps. Clutches comprise 20–25 eggs, 22 x 12mm.

KEELBACK SNAKES

Checkered Keelback Water Snake ■ *Xenochrophis piscator* 150cm
(*Assamese* Dhora; *Bengali* Jol Dhora; *Gujarati* Pani No Samp; *Hindi* Chappan Kuli; *Kannada* Neer Mandali; *Malayalam* Neer Koli; *Manipuri* Lilabob; *Marathi* Deward; *Nepali* Pani Samp; *Oriya* Pani Dhanda; *Punjabi* Dhobi Sapp; *Sindhi* Nadiwala; *Sinhalese* Diya Bariya, Diya Naya; *Tamil* Neerkkoli, Tanni Pambu, Tanni Saradi; *Telugu* Neella Pamu; *Urdu* Chittra Dhoobi Saamp)

DESCRIPTION Body stout; eyes have rounded pupils; nostrils directed slightly upwards; dorsal scales strongly keeled; dorsum olive-brown, with black spots arranged in 5–6 rows; head brown with black stripe from eye to upper lip, and from postoculars to edge of mouth. **DISTRIBUTION** India, Bangladesh, Bhutan, Nepal, Pakistan and Sri Lanka; range extends from Afghanistan to southern China and Southeast Asia. **HABITAT AND HABITS** Found in wetlands in plains, including flooded rice fields, ponds, marshes and rivers. Aquatic and active by day and night. Diet includes fish and frogs. Clutches comprise 30–80 eggs, 15–18 x 27–40mm. Incubation period 37–90 days. Hatchlings 110mm.

Triangled Keelback Water Snake
■ *Xenochropis trianguligerus* 120cm

DESCRIPTION Body slender; head large; eyes small; tail short; scales strongly keeled; dorsum blackish-brown, with orangish-red triangles on sides of neck and front of body, bright colours turning olive-brown or grey in older individuals; triangle-shaped dark marks on dorsum; lips cream coloured, with some scales on lips black edged; belly cream coloured. **DISTRIBUTION** Nicobar Islands, India; also Southeast Asia. **HABITAT AND HABITS** Found in streams in lowland rainforests; also in fields of rice paddies and vicinity of villages. Diet consists of frogs and their eggs, and tadpoles. Clutches comprise 5–15 eggs.

KEELBACK SNAKES/FALSE COBRAS

Andamans Keelback Water Snake ■ *Xenochrophis tytleri* 92cm
(*Hindi* Pani Samp)

DESCRIPTION Body stout, cylindrical; eyes have rounded pupils; nostrils directed slightly upwards; dorsal scales strongly keeled; dorsum olive-brown or tan, with 5 longitudinal dark brown stripes along body;

head buff-brown with black stripe from eye to upper lip, and from postoculars to edge of mouth, joining lowest dorsal stripe. DISTRIBUTION Andaman Islands, India. HABITAT AND HABITS Found in water bodies, especially near fields of rice paddies, and sluggish streams and marshes. Aquatic; sometimes also on land at night. Diet includes fish and frogs. Clutches comprise 8–87 eggs.

PSEUDOXENODONTIDAE (FALSE COBRAS)
This family contains six described species, all currently considered non-venomous. They are referred to as false cobras because in response to a threat, they raise the forebody and expand the skin in the neck region, in an effective mimicry of cobras.

Large-eyed False Cobra ■ *Pseudoxenodon macrops* 116cm
(*Nepali* Gorobi Samp)

DESCRIPTION Body stout; head distinct from neck; eyes large with rounded pupils; scales keeled; dorsum brownish-grey, red or olivaceous, with a series of yellow, reddish-brown or orange, dark-edged cross-bars or spots, and dorsolateral series of dark spots; nape has arrow-shaped dark marking; belly yellow, with large, quadrangular, black or dark brown spots or cross-bars. DISTRIBUTION Eastern and north-eastern India, Bangladesh, Bhutan and Nepal; also southern China and Southeast Asia. HABITAT

AND HABITS Found in middle-elevation forests. Terrestrial. Diet includes frogs and lizards. Oviparous, producing up to 10 eggs.

Elapid Snakes

ELAPIDAE (COBRAS, KRAITS, CORAL SNAKES AND SEA SNAKES)

The Elapidae are a family of venomous snakes found in tropical and subtropical regions of Asia, Australia, Africa, the Americas and Australasia. They are characterized by long, slender bodies; smooth scales; forehead with large shields; and eyes with rounded pupils. Elapid snakes have fixed fangs and their venom produces neurotoxic effects, including the disabling of muscle contraction and paralysis, especially of the respiratory system. Sea snakes are placed in this family, and have evolved laterally compressed bodies; reduced ventral scales; nostrils placed dorsally; paddle-like tails for swimming; and the ability to excrete salt and produce live young. A total of 361 living species is allocated to this family.

Indian Krait ■ *Bungarus caeruleus* 175cm

(*Bengali* Chitti Bora, Kalkeute, Kalochitti; *Gujarati* Kalotaro; *Hindi* Karayat; *Kannada* Godinagara Havu; *Malayalam* Valla Pambu, Yettadi Virian; *Marathi* Maniar; *Nepali* Karkat Nag; *Oriya* Chiti Shapa; *Punjabi* Chit Kaudiya, Kaudiya; *Sindhi* Pee-un, Sangchul; *Sinhalese* Magamaruwa, Thel Karawala; *Tamil* Kattu Viriyan, Ettadi Viriyan; *Telugu* Gaddi Parugudu, Katla Pamu; *Urdu* Kala Gandait, Sangchoor Saamp)

DESCRIPTION Body slender, triangular in cross-section, with raised vertebral region; head indistinct from neck; eyes small, pupils rounded; dorsals smooth; vertebral scales enlarged, hexagonal; dorsum steely-blue, black or dark brown, with 3–9 pale vertebral spots, followed by 38–56 narrow white bands, arranged in pairs, across body; upper lips cream; belly unpatterned cream. **DISTRIBUTION** India (except northeast), Bangladesh, Nepal, Pakistan and Sri Lanka; also Afghanistan. **HABITAT AND HABITS** Found in plains, in thinly wooded forests, agricultural fields and edges of human habitation. Nocturnal. Diet includes snakes, lizards, frogs and rodents. Clutches comprise 6–15 eggs, 35 x 19mm. Hatchlings 266–298mm. Venom highly neurotoxic and causes respiratory failure.

◾ ELAPID SNAKES ◾

Sri Lankan Krait ◾ *Bungarus ceylonicus* 135cm
(*Sinhalese* Dhunu Karawala, Polon Karawala; *Tamil* Yennai Viriyan, Yettadi Viriyan)

DESCRIPTION Body slender, triangular in cross-section, with raised vertebral region; head indistinct from neck; body cylindrical; eyes small; scales shiny; vertebral scales enlarged, hexagonal; dorsum steely-blue, black or dark brown, with 20 narrow white bands on body; bands indistinct or reduced to vertebral spots and disappear in adults; belly cream coloured, with dark bands. **DISTRIBUTION** Sri Lanka. **HABITAT AND HABITS** Widespread in secondary forests, agricultural fields and edges of human habitation, at up to 2,000m. Diet includes snakes, mice, geckos and skinks. Clutches comprise 4–10 eggs, 12–17 x 29–35mm. Hatchlings 230–278mm.

Banded Krait ◾ *Bungarus fasciatus* 225cm
(*Assamese* Gwala Hap; *Bengali* Shakhamukhi, Sankhini; *Hindi* Chit Kaudiya, Goman; *Manipuri* Linkhak; *Nepali* Ganguri Sarpa, Gun Gawari; *Oriya* Rona Shapa; *Telugu* Bungarum Pamu)

DESCRIPTION Body slender, triangular in cross-section, with raised vertebral region; tail short and stumpy; subcaudal scales undivided; forehead has 'V'-shaped mark; dorsum has alternating black-and-yellow bands that are subequal. **DISTRIBUTION** India (except extreme north, west and south), Bangladesh, Bhutan and Nepal; also Southeast Asia. **HABITAT AND HABITS** Found in lowlands, including forests, swamps and vicinity of villages. Diet includes water snakes, rat snakes, pythons, vine snakes, lizards, frogs and fish. Clutches comprise 4–14 eggs; incubation period 61 days. Hatchlings 250–300mm.

◾ ELAPID SNAKES ◾

Black Krait ◾ *Bungarus niger* 180cm

DESCRIPTION Body slender, triangular in cross-section, with raised vertebral region; head indistinct from neck; eyes rather small, pupils rounded; dorsals smooth; vertebral scales enlarged; last few subcaudal scales single; dorsum black or bluish-black; lips and throat cream coloured; belly and undersurface of tail cream coloured, with dark mottling. **DISTRIBUTION** Eastern Himalayas of eastern and north-eastern India, Bangladesh, Bhutan and Nepal. **HABITAT AND HABITS** Found in forests at low elevation and mid-hills, up to 1,490m. Diet and reproductive habits unknown.

Indian Coral Snake ◾ *Calliophis melanurus* 300mm
(*Sinhalese* Depath Kaluwa; *Tamil* Pavallap Pambu)

DESCRIPTION Body elongated, cylindrical, of equal thickness; neck indistinct from head; tail short; scales smooth, not shiny; black; pale spot behind parietals; back unpatterned; belly orange or yellow and no red on tail. **DISTRIBUTION** Peninsular India and Sri Lanka. **HABITAT AND HABITS** Found in lowland forests. Diet includes blind snakes. Reproductive habits unknown.

▪ Elapid Snakes ▪

Annulated Sea Snake ▪ *Hydrophis cyanocinctus* 189cm
(*Bengali* Chittul, Kalo Holud Lati Shap; *Hindi* Chittul; *Tamil* Kadal Sarai; *Urdu* Patta Samundri Saamp)

DESCRIPTION Body slender; head small, indistinct from neck, thickening towards posterior; dorsum olive or yellow, with darker transverse bands that may encircle body; belly yellowish-cream. **DISTRIBUTION** India, Bangladesh, Pakistan and Sri Lanka; **DISTRIBUTION** extends from Persian Gulf, to Southeast Asia and Japan. **HABITAT AND HABITS** Found in coastal waters. Diet includes fish, especially eels, and marine invertebrates. Clutches comprise 3–16 young, 381mm.

Hook-nosed Sea Snake ▪ *Hydrophis schistosus* 158cm
(*Bengali* Hooghly Patee; *Gujarati* Bambudiya, Daryano Samp; *Hindi* Darya Samp, Samudra Samp; *Malayalam* Valakadiyan Pambu; *Oriya* Dushta Shapa, Samudra Shapa; *Sinhalese* Valakkadia; *Tamil* Potai Pambu, Valaikadiyan; *Urdu* Chonchu Samundri Saamp)

DESCRIPTION Body stout; dorsals distinctly keeled; rostral scale on upper jaw overhung; dorsum of body and forehead greyish-olive, forehead darker, body with indistinct

darker markings; belly cream coloured in front, darkening to greenish-yellow towards tail. **DISTRIBUTION** India, Bangladesh, Pakistan and Sri Lanka. **DISTRIBUTION** extends from Persian Gulf, east to New Guinea and northern Australia. **HABITAT AND HABITS** Found in shallow coastal regions, including mangroves and some freshwater rivers. Diet consists of marine catfish. Ovoviviparous; clutches comprise 4–11 young; newborns 254–280mm.

ELAPID SNAKES

Yellow-lipped Sea Krait ■ *Laticauda colubrina* 171cm
(*Karen* Wu Te Paim We; *Onge* Tebule; *South Nicobarese* Jungliu, Vaugh Layu)

DESCRIPTION Body stout; tail flattened; pupils rounded; dorsum bluish-grey, with 36–50 black annuli separated by bluish-grey bands. **DISTRIBUTION** Andaman and Nicobar Islands, India, and a few isolated records from mainland south Asia; range of species extends east to South Pacific. **HABITAT AND HABITS** Found in tropical seas and ascends small islands to bask, rest and perhaps also digest food. Diet consists of eels. Eggs produced in caves on rocky islands. Clutches comprise 3–13 eggs, 45–92 x 20–31mm.

Monocled Cobra ■ *Naja kaouthia* 230cm
(*Assamese* Pheti Hap; *Bengali* Keute; *Manipuri* Kharou; *Oriya* Tampa Naga)

DESCRIPTION Body stout; head indistinct from body; neck capable of dilating into hood; dorsals smooth; dorsum brown, greyish-brown, blackish-brown or black; some individuals have darker bands; hood marking light circle with dark centre; pale throat colour extends less far backwards than in Spectacled Cobra (see p. 138). **DISTRIBUTION** Eastern to north-eastern India, Bangladesh, Bhutan and Nepal; also southern and eastern China, and Southeast Asia. **HABITAT AND HABITS** Found in forests and often encountered in agricultural fields and plantations. Diet includes rodents, frogs, fish and snakes. Clutches comprise 12–33 eggs. Incubation period 50–60 days.

ELAPID SNAKES

Spectacled Cobra ■ *Naja naja* 220cm
(*Bengali* Gokhura; *Gujarati* Asal Samp, Phanidhar; *Hindi* Bechashmi Nag; *Kannada* Nagara Havu; *Malayalam* Moorkan Sarpam; *Marathi* Nag; *Nepali* Gaoman Sarpa; *Oriya* Gokhura; *Punjabi* Chagli-wala, Kala Machhiar, Phaniar Sapp; *Rajasthani* Gogaji, Kala Hamp, Nagin; *Sindhi* Kala Nag, Taleehar, Pambu; *Sinhalese* Nagaya, Naia, Naya; *Tamil* Nalla Pambu; *Telugu* Thrachu Pamu; *Urdu* Shesh Nag)

DESCRIPTION Body stout; head indistinct from body; neck capable of dilating into hood; dorsals smooth; dorsum greyish-brown, brown-black or jet-black, without markings or with dark or light bands; hood marking light spectacle, which may occasionally be absent; pale throat colour extends further backwards than in Monocled Cobra (see p. 137). **DISTRIBUTION** Northern, peninsular and eastern India, Bangladesh, Bhutan, Nepal, Pakistan and Sri Lanka; also possibly Afghanistan. **HABITAT AND HABITS** Found in dry forests and plantations, occasionally entering human habitation. Diet includes rodents, frogs, birds and their eggs, lizards, fish and snakes. Clutches comprise 12–30 eggs. Incubation period about 60 days. Hatchlings 200–300mm.

Andaman Cobra ■ *Naja sagittifera* 64cm
(*Karen* Singwe-mwe, U Thaw Bleh)

DESCRIPTION Body stout; head indistinct from body; neck capable of dilating into hood; dorsals smooth; dorsum black in adults; juveniles have light chevron-shaped markings; monocle-like marking on hood, comprising pale circle with dark centre; throat and belly grey.

DISTRIBUTION Andaman Islands, India. **HABITAT AND HABITS** Found in forested areas and plantations. Diet and reproductive biology unknown, accepting rats in captivity.

ELAPID SNAKES

King Cobra ■ *Ophiophagus hannah* 5.5m
(*Assamese* Roja Pheti; *Bengali* Raj Gokhura, Shankhachur; *Hindi* Naga Raja, Nagraj, Rajnagj; *Kannada* Kalinagin, Kalinga Havu, Sarpa; *Karen* Wu Thaw, Khasia Bsein Yong; *Malayalam* Karamboda Pambu, Krishna Sarpam; *Manipuri* Ising Kharou; *Marathi* Rajsamp; *Nepali* Raj Gaoman; *Oriya* Ahiraj, Jhadakalua Shapa; *Tamil* Karanjati, Karru Nagam, Karinagam, Rajanagam; *Telugu* Kalinga Pamu, Naga Raju)

DESCRIPTION Body stout in adults, slender in juveniles; head distinct from neck, which is capable of dilating into elongated hood; dorsals smooth; large pair of occipital scales; dorsum dark brown, olive-brown or grey-black, with pale yellow or orange bands in young that may or may not persist in adults. **DISTRIBUTION** India, Bangladesh, Bhutan and Nepal; also southern and eastern China, and Southeast Asia. **HABITAT AND HABITS** Terrestrial as well as arboreal, sometimes being found on branches of tall trees. Diet includes snakes and occasionally monitor lizards. Females construct mound nest of leaves, where eggs are produced. Clutches comprise 24 eggs. Incubation period 63 days. Hatchlings 288–422mm.

MacClelland's Coral Snake ■ *Sinomicrurus macclellandi* 84cm
(*Bengali* Probal Shap)

DESCRIPTION Body slender, cylindrical; head short and rounded; dorsals smooth; subcaudals divided; dorsum reddish-brown, with 23–40 narrow, light-edged black stripes; tail has 2–6 black bands; belly yellowish-cream with black marks. **DISTRIBUTION** North-eastern India, Bangladesh and Nepal; also southern China and Southeast Asia. **HABITAT AND HABITS** Found in lowland forests and hills at 900–1,800m. Nocturnal. Diet includes snakes and lizards. Clutches comprise 6–14 eggs.

■ MUD SNAKES ■

HOMALOPSIDAE (MUD SNAKES)
The Homalopsidae, also known as mud snakes or Oriental-Australasian rear-fanged snakes, includes stout-bodied species that are predominantly freshwater inhabitants; a few are associated with coastal waters and estuaries, and a handful are terrestrial. The family comprises 53 living species.

Dog-faced Water Snake ■ *Cerberus rynchops* 127cm
(*Bengali* Gangmete, Jal-kata Shap, Jolbora, Khari Shap, Nona Bora, Nona Mete; *Gujarati* Kalu Dendu, Pani No Samp; *Hindi* Pani-ka-samp; *Marathi* Pansarp; *Oriya* Jalaganthia; *Sinhalese* Diya Bariya, Diya Polonga, Kuna Diya Kaluwa; *Tamil* Bokadam, Naithalayan, Upranar Pambu; *Telugu* Uppu Nella Pamu)

IDENTIFICATION Body moderately slender; head long, distinct from neck; eyes small, beady, dorsally placed, with rounded pupils; dorsals keeled; dorsum dark grey or olive, with

faint dark blotches and dark line along sides of head, across eyes; belly yellowish-cream, with dark grey areas. **DISTRIBUTION** India (including Andaman and Nicobar Islands), Bangladesh and Sri Lanka; also Southeast Asia. **HABITAT AND HABITS** Lives in mangrove mudflats; occasionally found in coastal rice fields, hiding in crab holes. Diet includes gobies, mudskippers, crabs and frogs. Ovoviviparous, producing 5–38 young, 150–200mm.

Common Smooth Water Snake ■ *Enhydris enhydris* 88cm
(*Assamese* Meni; *Bengali* Huria, Kanametuli, Paina Shap; *Nepali* Mach Giddhee; *Oriya* Kauchia Dhanda Sapo; *Tamil* Alai Pambu, Mutta Pambu; *Telugu* Ally Pam, Mutta Pam)

IDENTIFICATION Body stout; head relatively small, somewhat depressed, slightly distinct

from neck; snout rounded; nostrils situated on upper surface of head; pupils vertical; dorsals smooth; tail pointed; dorsum greyish-brown or olive-green, typically with dark vertebral and 2 light lateral stripes from upper surface of head to tail; belly yellowish-cream, with dark line along each side. **DISTRIBUTION** Eastern and north-eastern India, Bangladesh, Bhutan and Nepal; also southern China and Southeast Asia. **HABITAT AND HABITS** Found in fresh and brackish waters, such as slow-moving rivers, marshes, lakes and rice fields. Diet includes fish, frogs and more rarely lizards. Ovoviviparous, producing several clutches of 2–20 young, 155–206mm.

■ Mud Snakes ■

Siebold's Water Snake ■ *Enhydris sieboldii* 77cm

IDENTIFICATION Body stout; head relatively small, somewhat depressed, slightly distinct from body; snout rounded; nostrils situated on upper surface of head; eyes small, pupils rounded; dorsals smooth; tail moderate; dorsum comprises pattern of nearly 20 brown blotches with dark brown edges; belly grey, with dark line in middle. **DISTRIBUTION** Eastern and north-eastern India, Bangladesh, Bhutan and Nepal; also Myanmar. **HABITAT AND HABITS** Found in rivers, lakes and streams, and associated with muddy banks. Diet includes frogs and probably fish. Ovoviviparous, producing 5–7 young, 182mm.

White-bellied Mangrove Snake ■ *Fordonia leucobalia* 95cm

IDENTIFICATION Body stout; head short, wide, slightly wider than neck; head scales large and distinct; loreal absent; lower jaw short; dorsals smooth; dorsum dark grey or brown, with light spots, or light grey, yellow or orange with dark spots; belly cream coloured, sometimes with small dark spots; lips yellowish-cream. **DISTRIBUTION** Eastern India and Andaman Islands and Bangladesh; also Southeast Asia, New Guinea and northern Australia. **HABITAT AND HABITS** Found in tidal rivers. Crabs form main component of diet, though small fish are also consumed. Ovoviviparous, producing 2–17 young, 176–196mm.

Mud Snakes/Slug-eating Snakes

Glossy Marsh Snake ■ *Gerarda prevostiana* 53cm
(*Gujarati Pani No Samp*)

IDENTIFICATION Body slender, cylindrical; head distinct from neck; eyes small, with vertically elliptical pupils; dorsals smooth; tail short and pointed; dorsum unpatterned grey or brown; lips and lower scales of dorsum cream; belly brownish-cream, with median dark streaks. **DISTRIBUTION** West and east coasts of India, Bangladesh and Sri Lanka; also Southeast Asia. **HABITAT AND HABITS** Found in mangrove swamps. Diet comprises recently moulted crabs, whose limbs are torn off and consumed; fish and shrimps also eaten. Clutches of 5 are produced.

PAREIDAE (Slug-eating Snakes)
The Pareatidae are a small family (20 described species) of non-venomous snake. A majority of the species are snail or slug eaters. They are characterized by a short and high head, large eyes, large genials (chin shields), lack of a mental groove, solid teeth on both jaws and enormous nasal glands. They are nocturnal, arboreal and oviparous.

Spotted Slug-eating Snake ■ *Pareas macularius* 45cm

DESCRIPTION Body slender, laterally compressed; head short, rounded, distinct from neck; snout short; eyes large, pupils vertical; tail short; vertebrals not enlarged; median 3–7 dorsals keeled; dorsum light or dark grey, with dark spots or irregular black cross-bars; lips pale; pale nuchal collar occasionally present; belly cream coloured, with irregular black speckles. **DISTRIBUTION** Eastern and North-east India and Bangladesh; also Southeast Asia. **HABITAT AND HABITS** Found in subtropical forests. Associated with low vegetation. Diet consists of slugs and snails. Clutches comprise up to 6 eggs.

Slug-eating Snakes/Vipers & Pit Vipers

Montane Slug-eating Snake ■ *Pareas monticolus* 61cm
(*Bengali* Shamuk-khor Shap)

DESCRIPTION Body slender, laterally compressed; head short, rounded, distinct from neck; snout short; eyes large, pupils vertical; tail short; vertebral rows enlarged; dorsals weakly keeled; dorsum mid-brown, with vertical blackish-brown bars on flanks; black postocular stripe to nape; black streak from eye to angle of jaws; forehead brown with black spots; belly yellow with brown spots. **DISTRIBUTION** Eastern and north-eastern India and Bangladesh; also Southeast Asia. **HABITAT AND HABITS** Found in evergreen hill forests. Associated with low vegetation. Diet includes snails and slugs. Clutches comprise up to 8 eggs. Hatchlings 168–178mm.

VIPERIDAE (Vipers and Pit Vipers)
The Viperidae family contains the venomous vipers that are nearly cosmopolitan in distribution, with 339 described species. They are identifiable by their large heads that are distinct from their narrow necks; hinged, hollow fangs; mostly keeled scales; stout bodies (except in the arboreal species); and short, prehensile tails. The pit vipers have sensory pits on their snouts that act as thermal detectors for locating warm-bodied prey such as rats and birds. Most are sit-and-wait predators of warm-blooded prey and tend to be ovoviviparous, though oviparity also exists among them. Viper bites tend to produce haemotoxic effects, including pain, local swelling, necrosis and coagulopathy.

VIPERS & PIT VIPERS

White-lipped Pit Viper ■ *Cryptelytrops albolabris* 104cm
(*Manipuri* Naril)

DESCRIPTION Body slender; head long, distinct from neck; small scales on forehead; dorsals keeled; tail prehensile; iris yellow; head and dorsum green; males have white stripe on first row of dorsals, which is indistinct or absent in females; belly green or yellowish-cream. **DISTRIBUTION** North-eastern India and Bangladesh; also southern China and Southeast Asia. **HABITAT AND HABITS** Found in forests, near streams. Diet comprises mice, birds, lizards and frogs. Ovoviviparous, producing 7–16 young. Neonates 120–180mm.

Anderson's Pit Viper ■ *Cryptelytrops andersoni* 110cm
(*Hindi* Andha Samp; *Karen* Jungli Mbuin)

DESCRIPTION Body slender; head distinct from neck; snout long; scales on forehead not tuberculate; dorsals strongly keeled; dorsum with brown variegation on olive ground colour; belly pale brown. **DISTRIBUTION** Andaman Islands, India. **HABITAT AND HABITS** Found in lowland rainforests, as well as clearings within villages and edges of agricultural fields. Arboreal. Diet comprises rats and geckos. Reproductive habits unknown.

VIPERS & PIT VIPERS

Cantor's Pit Viper ■ *Cryptelytrops cantori* 115cm

DESCRIPTION Body stout; head distinct from neck; snout long; forehead scales smooth or weakly keeled; dorsals smooth or weakly keeled; dorsum olive, or light or dark brown, with or without darker brown or yellow spots; pale stripe along eyes; another pale stripe along flanks; belly cream or green; undersurface of tail brown spotted. **DISTRIBUTION** Kamorta Island, Central Nicobars, India. **HABITAT AND HABITS** Found in lowland sites, including vicinity of human habitation. Diet unknown, and presumably consists of birds and small mammals. Reproductive habits unknown.

Red-tailed Pit Viper ■ *Cryptelytrops erythrurus* 105cm

DESCRIPTION Body slender; head long, narrow, distinct from neck; body stout; small scales on forehead; forehead scales keeled; dorsals strongly keeled; tail prehensile; iris yellow; dorsum bright green; belly pale green or yellow; tip of tail spotted or mottled with brown. **DISTRIBUTION** North-eastern India, Bangladesh, Bhutan and Nepal; also Southeast Asia. **HABITAT AND HABITS** Found in forests, generally in vicinity of streams. Diet comprises rodents and birds. Reproductive habits unknown.

VIPERS & PIT VIPERS

Northern Pit Viper ■ *Cryptelytrops septentrionalis* 60cm

DESCRIPTION Body slender; head relatively long, narrower than neck; small scales on forehead; dorsals strongly keeled; tail prehensile; eyes yellow; dorsum green; belly pale green or yellow; upper lips pale green; pale yellow stripe runs from below loreal pit to back of head; tip of tail brown. DISTRIBUTION Northwestern India and Nepal. HABITAT AND HABITS Found in submontane forests, generally in vicinity of streams. Diet comprises rodents, lizards and birds. Reproductive habits unknown.

Russell's Viper ■ *Daboia russelii* 185cm

(*Bengali* Chandra Bora; *Gujarati* Chital; *Hindi* Dabua, Ghonus; *Kannada* Balikadak Havu; *Malayalam* Adarsha Mandali Ruthamandali, Ruthram Mandali; *Manipuri* Lindu Mayaknaibi; *Marathi* Ghonas; *Nepali* Ghoda Karet; *Oriya* Chandan Boda Shapa; Chandra Bora; *Punjabi* Kaudian-wala Sapp; *Sindhi* Khuppur, Korail; *Sinhalese* Dhara Polonga, Tic/Tith Polonga; *Tamil* Kanardi Viriyan; *Telugu* Katuka Rekula Poda; *Urdu* Koriala Afi)

DESCRIPTION Body stout; head large, neck thin; tail short; pupils vertical; forehead with small scales; nostrils large; dorsals keeled; dorsum brown, with 3 rows of spots along body, a dark brown one along midline, and a blackish-brown or black one on each side; belly cream coloured, with numerous small, crescentic marks on belly scales. DISTRIBUTION India, Bangladesh, Nepal, Pakistan and Sri Lanka. HABITAT AND HABITS Inhabits grassland, scrub forests and other open forests, entering agricultural fields with tall grass. Diet comprises rodents, crabs, frogs, lizards and birds. Ovoviviparous, producing clutches of 6–40 young (exceptionally 63).

◼ Vipers & Pit Vipers ◼

Saw-scaled Viper ◼ *Echis carinatus* 80cm
(*Baluchi* Phissi; *Bengali* Bankaraj; *Gujarati* Padaku, Phoorsa; *Hindi* Afai, Phoorsa; *Kannada* Kallu Havu; *Malayalam* Churatta, Surtai; *Marathi* Phoorsa, Phoorsi; *Oriya* Dhuli Naga; *Punjabi* Jalebi, Khappra; *Sindhi* Jalabia, Khuppar, Loon Dee; *Sinhalese* Vali, Weli Polonga; *Tamil* Suruttai, Suratti Pambu; *Urdu* Gunas, Khappra Saamp)

DESCRIPTION Body moderately stout; head oval, slightly distinct from neck; dorsals with sharp keels; eyes large; dorsum brown to greyish-brown, with series of pale, dark-edged blotches that are fused on each side to chevron-like marks; belly cream coloured. **DISTRIBUTION** India (except wet zone of south and north-east), Pakistan and Sri Lanka; range extends westwards to the Middle East. **HABITAT AND HABITS** Found in arid plains and deserts of peninsula and areas to its west. Diet includes mice, squirrels, lizards, snakes, frogs, locusts and centipedes. Clutches comprise 3–19 live young, 9–13mm. Northern subspecies (*multisquamatus*; 80cm) much larger than the one from peninsular India (*carinatus*; 30cm).

From Rajasthan

From Sri Lanka

Leaf-nosed Viper ◼ *Eristicophis macmahoni* 66cm
(*Urdu* Titli Afi)

DESCRIPTION Body stout; head distinct from body; forehead scales small, keeled; lateral keels arranged in oblique series; ventrals with strong lateral keel; tail prehensile; dorsum greyish-yellow, with small black spots; spots on flanks distinct; dark subocular streak from eye to corner of mouth; belly cream coloured. **DISTRIBUTION** Balochistan, Pakistan; also Afghanistan and eastern Iran. **HABITAT AND HABITS** Found in deserts. Associated with fine sand with sparse vegetation, including grass and bushes. Diet consists of arthropods and lizards. Clutches comprise 8–10 eggs, 24–34 x 14–18mm.

◾ Vipers & Pit Vipers ◾

Himalayan Pit Viper ◾ *Gloydius himalayanus* 86cm
(*Kashmiri* Pohur; *Urdu* Hafra Afi)

DESCRIPTION Body relatively stout; head somewhat flattened, distinct from neck, with enlarged scales; deep sensory pits between eye and nostril; dorsum brown, with darker patterns, typically with 23–45 cross-bars demarcated by their darker edges. DISTRIBUTION Northern India, Bhutan, Nepal and Pakistan; also China and possibly Afghanistan. HABITAT AND HABITS Found in low to middle elevations of the Himalayas at 1,524–4,877m. Diet comprises rodents, skinks and centipedes. Ovoviviparous, producing 3–7 young, 102–160mm.

Hump-nosed Pit Viper ◾ *Hypnale hypnale* 55cm
(*Sinhalese* Polon Thelissa, Geta Polonga, Kuna Katuwa; *Tamil* Kopi Viriyan, Kuzhi Viriyan)

DESCRIPTION Body stout; head distinct from neck; snout upturned; pupils vertical; dorsum light brown to blackish-brown, stippled with grey and brown; series of 20–33 oval or triangular marks on flanks, meeting at vertebral region; dark stripe from eye to jaws; dark cross-bands on tail. DISTRIBUTION Western Ghats of India and central Sri Lanka. HABITAT AND HABITS Found in lowland forests and associated with leaf litter, tree buttresses, rocks and low vegetation, at elevations of 30–1,500m. Diet includes geckos and rodents. Ovoviviparous, producing 4–17 young, 130–145mm.

Vipers & Pit Vipers

Millard's Hump-nosed Pit Viper ■ *Hypnale nepa* 39cm
(*Sinhalese* Mukalan Thelissa, Mukalan Kuna Katuwa; *Tamil* Viriyan Pambu)

DESCRIPTION Body stout; snout distinctly upturned, with wart-like projection on tip, consisting of 7–14 tiny scales; pupils vertical; dorsum light to dark brown, or pale olive, flecked and mottled with darker areas, pattern consisting of 17–26 suboval or subtriangular brown blotches; forehead brown; pair of broad dark stripes from back of head to neck; another broad dark stripe from upper lip to sides of throat and neck.
DISTRIBUTION Sri Lanka.
HABITAT AND HABITS Found in submontane and montane forests at 914–2,000m, and sometimes occurs in plantations. Diet includes snakes, skinks and frogs. Ovoviviparous, producing 4–6 live young.

Blotched Pit Viper ■ *Ovophis monticola* 115cm
(*Khasia* Bsein Longkru; *Manipuri* Tinducharang; *Nepali* Pahari Harew)

DESCRIPTION Body stout; head broad, distinct from neck; eyes relatively small, pupils vertical; dorsals smooth; tail not prehensile; dorsum brownish-grey or yellowish-pink, with dark brown blotches; forehead darker.
DISTRIBUTION Eastern Himalayas of eastern and north-eastern India, Bangladesh, Bhutan and Nepal; also southern China and Southeast Asia.
HABITAT AND HABITS Found in montane forests. Diet includes rats and mice. Oviparous; clutches comprise 5–18 eggs, 18–20mm.

VIPERS & PIT VIPERS

Pope's Pit Viper ■ *Popeia popeiorum* 100cm

DESCRIPTION Body slender; head small and flat, distinct from neck; pupils vertically elliptical; tail prehensile; scales on forehead and sides of head smooth; dorsals smooth or weakly keeled; dorsum bright green or pale green; pale line along flanks and tail; eyes red; males have red postocular stripe and red ventrolateral stripe; belly light green; tail reddish-brown. **DISTRIBUTION** Eastern Himalayas of eastern and north-eastern India, Bhutan and Bangladesh; also Southeast Asia.

HABITAT AND HABITS Inhabits montane and submontane forests. Diet consists of birds, frogs, lizards, rats and squirrels. Ovoviviparous, producing clutches of 10 young. Neonates 178mm.

Jerdon's Pit Viper ■ *Protobothrops jerdonii* 99cm

DESCRIPTION Body slender; snout relatively long; forehead scales reduced; dorsals strongly keeled; dorsum greenish-yellow or olive, with series of reddish-brown blotches, edged with black; forehead black, with fine yellow lines; upper lip yellow; belly anteriorly cream, blotched with black from midbelly to tail. **DISTRIBUTION** North-eastern India, Bangladesh and Bhutan; also southern China and Southeast Asia. **HABITAT AND HABITS** Found in montane evergreen forests and edges of agricultural areas. Diet presumably consists of small rodents and birds. Ovoviviparous, producing clutches of 4–9 young.

◾ Vipers & Pit Vipers ◾

Brown-spotted Pit Viper ◾ *Protobothrops mucrosquamatus* 116cm
(*Mishing* Borta-ta-sang)

DESCRIPTION Body slender; head relatively elongate; forehead scales reduced; dorsals strongly keeled; dorsum greyish-olive or brown, with series of large brown spots with dark edges; head sometimes has dark streak on sides; tail light brown with black spots; belly cream with light brown speckles. **DISTRIBUTION** North-eastern India and Bangladesh; also southern China and Southeast Asia. **HABITAT AND HABITS** Found in montane evergreen forests. Diet comprises frogs, lizards, bats, rats and birds. Oviparous, producing clutches of 5–13 eggs, 35 x 20mm.

Bamboo Pit Viper ◾ *Trimeresurus gramineus* 80cm
(*Bengali* Bansh Bora, Gechho Bora; *Gujarati* Kamdia, Leelo Samp, Nagubiyo; *Hindi* Hara Gonas; *Konkani* Chapdi; *Malayalam* Chattithalayan, Kattu Mandli; *Marathi* Hara Gonas; *Oriya* Garta-mastaki Baunsa Boda, Gendamundia Boda; *Tamil* Pachchai Viriyan, Pul Viriyan; *Telugu* Bodroo Pam)

DESCRIPTION Body slender; head long, narrow, distinct from neck; forehead scales small; dorsals weakly keeled; tail prehensile; eyes yellow; dorsum green or greenish-yellow, sometimes with dark brown blotches; upper lips cream coloured; dark postocular stripe; belly pale green or yellowish-cream. **DISTRIBUTION** Peninsular India. **HABITAT AND HABITS** Found in forested habitats, including bamboo groves. Arboreal. Diet comprises lizards and birds. Ovoviviparous, producing clutches of 6–20 young. Neonates 120mm.

■ Vipers & Pit Vipers ■

Sri Lankan Green Pit Viper ■ *Trimeresurus trigonocephalus* 120cm
(*Sinhalese* Pala Polonga; *Tamil* Kopi Viriyan, Pachchai Viriyan)

DESCRIPTION Body stout, head large, triangular; neck distinct; forehead scales large; tail short, prehensile; dorsum green with black variegated pattern; black temporal stripe present; vertebral area has tinge of yellow; tail black; belly light greenish-yellow or grey. **DISTRIBUTION** Sri Lanka. **HABITAT AND HABITS** Found in lowland evergreen forests. Arboreal species from grassland and rainforests, and plantations of cardamom, cocoa, coffee and tea, at 153–1,075m elevation. Diet includes frogs, rodents and birds. Ovoviviparous, producing 5–26 young. Neonates 200–250mm.

Medo Pit Viper ■ *Viridovipera medoensis* 56cm

DESCRIPTION Body slender; head flattened, triangular, distinct from neck; scales on upper snout enlarged; scales have obtuse keels on vertebral region and sides; dorsum dark green, sometimes edged with turquoise-blue, with bicoloured white/red ventrolateral stripe; belly pale green. **DISTRIBUTION** Northeastern India, China and Southeast Asia. **HABITAT AND HABITS** Found in tropical and subtropical montane forests at 500–1,400m. Arboreal; associated with bamboo forests and vicinity of streams. Diet comprises frogs and rodents. Reproductive biology unknown.

VIPERS & PIT VIPERS/BLIND SNAKES

Stejneger's Pit Viper ■ *Viridovipera stejnegeri* 112cm

DESCRIPTION Body slender; head large, distinct from neck; forehead scales smooth; dorsals keeled; dorsum and head bright green; belly pale or greenish-cream; eye reddish-brown; males have white postocular stripe; another stripe, bordered at bottom with red or orange, runs along flanks. **DISTRIBUTION** North-east India; also Southeast Asia and southern China. **HABITAT AND HABITS** Inhabits hill forests at up to 2,845m. Arboreal. Diet comprises rodents, birds, lizards and frogs. Ovoviviparous, producing 3–10 young.

TYPHLOPIDAE (BLIND SNAKES)
The Typhlopidae can be recognized by their mostly vestigial eyes; teeth in the lower jaws; tail ending in a horn-like scale; and a rostral scale overhanging the mouth to form a shovel-like burrowing structure. They are fossorial, and most feed on ants, termites and worms. To date, 267 species have been described.

Large Blind Snake ■ *Argyrophis diardii* 43mm
(*Assamese* Donga; *Bengali* Telia Saap; *Manipuri* Tipunnapunn; *Urdu* Moota Kainchwa Saamp)

DESCRIPTION Body stout; head indistinct and blunt; eyes visible; tail pointed; dorsum dark brown, each scale with indistinct light transverse streak; belly and flanks light brown. **DISTRIBUTION** Northern and north-eastern India, Bangladesh, Nepal and Pakistan; also Southeast Asia. **HABITAT AND HABITS** Inhabits lowland evergreen forests. Fossorial; associated with soil, rotting logs and underneath of debris. Diet comprises insects and earthworms. Either ovoviviparous or oviparous, producing up to 14 young.

BLIND SNAKES

Brahminy Blind Snake ■ *Indotyphlops braminus* 175mm
(*Assamese* Kshantia Hap; *Bengali* Kecho Shap, Puiya Shap; *Gujarati* Bandhani Chakam; *Hindi* Andha Samp, Do-mukh Ka Samp; *Malayalam* Chevi Ambu, Kurudi Pambu; *Marathi* Dauav Kadu; *Nepali* Andha Samp, Neliya Sarp; *Oriya* Do-mundia Shapa, Telia Shapa; *Punjabi* Gandoa Sapp; *Rajasthani* Kana; *Sinhalese* Dumuta Kanaulla; *Tamil* Pooran, Sevittu Pambu, Manallai Pambu, Seer Pambu, Sevi Pambu; *Telugu* Chevi Pamu, Guddi Pamu; *Urdu* Sapolia)

DESCRIPTION Body slender; head rounded; nostrils lateral; eyes distinct; head scales larger than dorsals; tail terminates in spine; dorsum black or brown; belly lighter;

snout and tip of tail paler. **DISTRIBUTION** India, Bangladesh, Bhutan, Nepal, Pakistan and Sri Lanka; tropical and subtropical regions of Asia, Central and North America, Australia, Africa and islands in Indian and Pacific Oceans. **HABITAT AND HABITS** Found in forests as well as disturbed areas. Diet comprises termites, ants, their larvae and earthworms. A parthenogenetic, oviparous snake.

Jerdon's Blind Snake ■ *Indotyphlops jerdoni* 28cm
DESCRIPTION Body slender; snout rounded; eyes distinct; tail bluntly pointed, ending in cream-coloured spine; dark dorsum brown to nearly black; belly light brown; snout

and anal region cream coloured. **DISTRIBUTION** Eastern and north-eastern India, Bangladesh, Bhutan and Nepal; also Southeast Asia. **HABITAT AND HABITS** Inhabits mid-elevation evergreen forests. Fossorial; associated with soil, underneaths of rocks and insides of dead trees. Diet comprises termites and earthworms. Reproductive habits unknown.

CROCODILES

CROCODYLIDAE (CROCODILES)
Crocodiles are large aquatic predators. Adaptations for an aquatic life include a streamlined body, webbed feet, a palatal flap (tissue at the back of the mouth that blocks the entry of water), and the ability to close the nostrils during submergence. A total of 16 living species represent the family today. The family contains some of the largest living reptiles. Crocodilians are more closely related to birds and dinosaurs than to most other reptiles, having a cerebral cortex and a four-chambered heart. The Indian subcontinent is home to two species of true crocodile, one from saltwater habitats, the other from fresh water. They are predators of small to large prey, and large crocodiles are known to attack humans. The sex of crocodiles is determined by egg incubation temperature.

Mugger Crocodile ■ *Crocodylus palustris* 5m
(*Bengali* Kumir; *Gujarati* Mugger; *Hindi* Mugger-mach; *Malayalam* Cheengkani; *Nepali* Gohee, Magar Gohee; *Oriya* Gomuhan Kumbhira, Kuji Kumbhira, Matia Kumbhira; *Sindhi* Wugu; *Sinhalese* Ala Kimbula, Hale Kimbula; *Tamil* Kulathi Muthalai, Muthalai; *Urdu* Baghori, Magar Machh)

DESCRIPTION Snout relatively broad and heavy; forehead concave; ridges in front of eyes absent; dorsal scales in 16–17 rows on trunk; postoccipital scutes absent; 13–14 pairs of teeth on upper jaw and 14–15 pairs on lower jaw; juveniles light tan or brown with dark cross-bands on body and tail; adults grey to brown, usually without dark bands.

DISTRIBUTION India, Pakistan, Nepal, Bangladesh and Sri Lanka. Also Iran. **HABITAT AND HABITS** Found in fresh water, including rivers, lakes, dams and reservoirs, generally away from tidal influence. Insects and small vertebrates, including fish and frogs, are consumed by juveniles, while adults eat mammals as large as deer and goats, as well as small mammals, water birds, fish, snakes, lizards and turtles; they also feed on carrion. Nest is a hole in the ground, where 10–50 eggs are laid, and 2 clutches of eggs may be produced annually. Eggs 64–84 x 40–51mm.

CROCODILES/GHARIALS

Saltwater Crocodile ■ *Crocodylus porosus* 6.2m
(*Bengali* Nona-joler Kumir; *Oriya* Baula Kumbhira, *Sinhalese* Gatte Kimbula, *Shompen* Neaw Diuk; *South Nicobarese* Ko-on; *Tamil* Semmukhan Muthalai)

DESCRIPTION Head large; snout heavy; paired ridges from orbit to centre of snout; scales on back oval; juveniles brightly coloured, black spotted or blotched on pale yellow or grey background; colouration less bright in adults. **DISTRIBUTION** Sunderbans and Bhitarkanika mangroves on east coast, as well as Andaman and Nicobar Islands in India, coastal Bangladesh and Sri Lanka. Range includes the Seychelles, Indo-Malaya, to New Guinea, the Philippines, Australia and islands in the South Pacific. **HABITAT AND HABITS** Associated with mangrove forests, estuaries and sea coasts, as well as tidal rivers and lakes in their vicinity. Diet of juveniles includes crabs, shrimps, insects, fish, lizards and snakes, while adults eat turtles, birds and mammals. Occasionally attacks humans. Constructs a mound nest, in which 60–80 eggs are deposited, and females guard nest.

GAVIALIIDAE (GHARIALS)
The Indian Gharial, and sometimes the Sunda Gharial, are placed in this family. The Indian species is large with a narrow snout, adaptive for its diet of fish. It has 110 sharp, interdigitated teeth, which are adaptations for its specialized diet. Adult males have a distinctive swelling at the tip of the snout.

Indian Gharial ■ *Gavialis gangeticus* 7m
(*Assamese* Godul; *Bengali* Gharial, Mecho Kumhir; *Hindi* Bahsoolia Nakar, Gharial, *Nepali* Chimpta Gohee, *Oriya* Ghadiala Naka, *Sindhi* Say-sar; *Urdu* Sassar)

DESCRIPTION Body relatively slender, elongated; snout slender, parallel sided, with distinctive knob in adult males; about 100 sharp, interlocking teeth; dorsum olive to tan,

with dark blotches or bands on dorsum; belly pale. **DISTRIBUTION** Rivers Indus, Ganga, Brahmaputra and Mahanadi of northern and eastern India, Pakistan, Nepal and Bangladesh; extinct in Bhutan. **HABITAT AND HABITS** Inhabits large rivers with sand banks, and is a specialized fish eater. Nests on sand banks or alluvial deposits, laying 7–60 eggs, 59–64mm. Incubation period 62 days. Hatchlings 172–176mm.

CHECKLIST OF THE REPTILES OF INDIA

IUCN 2017 Red List Status (version 2016–13)

LC   Least Concern
LR / NT   Lower Risk / Near Threatened
VU   Vulnerable
EN   Endangered
CR   Critically Endangered
EX   Extinct in the Wild
NE   Not Evaluated

ORDER TESTUDINES

Testudinidae (Land Tortoises)

Indian Star Tortoise *Geochelone elegans* VU
Elongated Tortoise *Indotestudo elongata* EN
Travancore Tortoise *Indotestudo travancorica* VU
Asian Giant Tortoise *Manouria emys* EN
Central Asian Tortoise *Testudo horsfieldii* VU

Geoemydidae (Pond Turtles)

River Terrapin *Batagur baska* CR
Three-striped Roofed Turtle *Batagur dhongoka* EN
Painted Roofed Turtle *Batagur kachuga* CR
Malayan Box Turtle *Cuora amboinensis* VU
Keeled Box Turtle *Cuora mouhotii* EN
Gemel's Leaf Turtle *Cyclemys gemeli* NE
Spotted Pond Turtle *Geoclemys hamiltonii* VU
Crowned River Turtle *Hardella thurjii* VU
Arakan Hill Turtle *Heosemys depressa* CR
Tricarinate Hill Turtle *Melanochelys tricarinata* VU
Indian Black Turtle *Melanochelys trijuga* LR/NT
Indian Eyed Turtle *Morenia petersi* VU
Brown Roofed Turtle *Pangshura smithii* LR/NT
Assam Roofed Turtle *Pangshura sylhetensis* EN
Indian Roofed Turtle *Pangshura tectum* LR/NT
Indian Tent Turtle *Pangshura tentoria* LR/NT
Forest Cane Turtle *Vijayachelys silvatica* EN

Trionychidae (Softshell Turtles)

Malayan Softshell Turtle *Amyda cartilaginea* VU
Narrow-headed Softshell Turtle *Chitra indica* EN
Sri Lankan Flapshell Turtle *Lissemys ceylonensis* NE
Indian Flapshell Turtle *Lissemys punctata* LR/NT
Indian Softshell Turtle *Nilssonia gangeticus* VU
Indian Peacock Softshell Turtle *Nilssonia hurum* VU
Leith's Softshell Turtle *Nilssonia leithii* VU
Black Softshell Turtle *Nilssonia nigricans* EX (wild populations found since assessment)
Asian Giant Softshell Turtle *Pelochelys cantorii* EN

Cheloniidae (Marine Turtles)

Loggerhead Sea Turtle *Caretta caretta* VU
Green Turtle *Chelonia mydas* EN
Hawksbill Sea Turtle *Eretmochelys imbricata* CR
Olive Ridley Sea Turtle *Lepidochelys olivacea* VU

Dermochelyidae (Leatherback Sea Turtle)

Leatherback Sea Turtle *Dermochelys coriacea* VU

ORDER SQUAMATA

Agamidae (Agamid Lizards)

Green Crested Lizard *Bronchocela cristatella* NE
Nicobar Crested Lizard *Bronchocela danieli* NE

Maned Forest Lizard *Bronchocela jubata* LC

Camorta Forest Lizard *Bronchocela rubrigularis* NE

Laungwala Toad-headed Lizard *Bufoniceps laungwalansis* NE

Yellow-lipped Forest Lizard *Calotes aurantilabium* DD

Bhutan Forest Lizard *Calotes bhutanensis* NE

Green Forest Lizard *Calotes calotes* NE

Painted-lipped Lizard *Calotes ceylonensis* NE

de Silva's Forest Lizard *Calotes desilvai* NE

Elliot's Forest Lizard *Calotes ellioti* LC

Emma Gray's Forest Lizard *Calotes emma* NE

Large-scaled Forest Lizard *Calotes grandisquamis* LC

Jerdon's Forest Lizard *Calotes jerdoni* NE

Crestless Lizard *Calotes liocephalus* EN

Whistling Lizard *Calotes liolepis* NE

Kelum's Forest Lizard *Calotes manamendrai* NE

Maria's Lizard *Calotes maria* NE

Lesser Ground Lizard *Calotes minor* DD

Moustached Forest Lizard *Calotes mystaceus* NE

Nilgiri Forest Lizard *Calotes nemoricola* LC

Black-lipped Lizard *Calotes nigrilabris* NE

Rohan's Forest Lizard *Calotes pethiyagodai* NE

Roux's Forest Lizard *Calotes rouxii* LC

Garden Lizard *Calotes versicolor* NE

Rough-horned Lizard *Ceratophora aspera* VU

Erdelen's Horned Lizard *Ceratophora erdeleni* NE

Karu's Horned Lizard *Ceratophora karu* NE

Rhinoceros-horned Lizard *Ceratophora stoddartii* NE

Leaf-nosed Lizard *Ceratophora tennentii* EN

Pygmy Lizard *Cophotis ceylanica* NE

Knuckles Pygmy Lizard *Cophotis dumbara* CR

Andamans Short-tailed Lizard *Coryphophylax brevicaudus* NE

Bay Islands Forest Lizard *Coryphophylax subcristatus* NE

Blanford's Flying Lizard *Draco blanfordii* NE

Western Ghats Flying Lizard *Draco dussumieri* LC

Spotted Flying Lizard *Draco maculatus* LC

Norville's Flying Lizard *Draco norvilii* NE

Anderson's Mountain Lizard *Japalura andersoniana* NE

Das' Mountain Lizard *Japalura dasi* DD

Kumaon Mountain Lizard *Japalura kumaonensis* NE

Large Mountain Lizard *Japalura major* NE

Ota's Mountain Lizard *Japalura otai* NE

Flat-backed Mountain Lizard *Japalura planidorsata* NE

Burmese Mountain Lizard *Japalura sagittifera* NE

Three-keeled Mountain Lizard *Japalura tricarinata* LC

Variegated Mountain Lizard *Japalura variegata* LC

Agror Rock Agama *Laudakia agrorensis* NE

Day's Rock Agama *Laudakia dayana* NE

Black-tailed Rock Agama *Laudakia melanura* NE

Large-scaled Rock Agama *Laudakia nupta* NE

Nuristan Rock Agama *Laudakia nutistanica* NE

Pakistani Rock Agama *Laudakia pakistanica* NE

Kashmiri Rock Agama *Laudakia tuberculata* NE

Hump-nosed Lizard *Lyriocephalus scutatus* NT

Indian Kangaroo Lizard *Otocryptis beddomii* EN

Black-spotted Kangaroo Lizard *Otocryptis nigristigma* NE

Sri Lankan Kangaroo Lizard *Otocryptis wiegmanni* NE

Caucasian Agama *Paralaudakia caucasia* NE

Himalayan Agama *Paralaudakia himalayana* NE

Small-scaled Agama *Paralaudakia microlepis* NE

Clarks' Toad-headed Agama *Phrynocephalus clarkorum* NE

Alcock's Toad-headed Agama *Phrynocephalus euptilopus* NE

Yellow-speckled Toad Agama *Phrynocephalus luteoguttatus* NE

Black-tailed Toad Agama *Phrynocephalus maculatus* NE

Ornate Toad Agama *Phrynocephalus ornatus* NE

Grey Toad Agama *Phrynocephalus scutellatus* NE

Theobald's Toad-headed Agama *Phrynocephalus theobaldi* LC

Blanford's Rock Agama *Psammophilus blanfordanus* LC

South Indian Rock Agama *Psammophilus dorsalis* LC

Andamans False Blood-sucker *Pseudocalotes andamanensis* NE

Godwin-Austin's False Blood-sucker *Pseudocalotes austeniana* NE

Kingdon-Ward's False Blood-sucker *Pseudocalotes kingdonwardi* NE

Small False Blood-sucker *Oriocalotes paulus* NE

Green Fan-throated Lizard *Ptyctolaemus gularis* NE

Hardwicke's Spiny-tailed Lizard *Saara hardwickii* NE

Anaimalai Spiny Lizard *Salea anamallayana* LC

Horsfield's Spiny Lizard *Salea horsfieldii* LC

Darwin's Fan-throated Lizard *Sarada darwini* NE

Deccan Fan-throated Lizard *Sarada deccanensis* NE

Superb Fan-throated Lizard *Sarada superba* NE

Sri Lankan Fan-throated Lizard *Sitana bahiri* NE

Devaka's Fan-throated Lizard *Sitana devakai* NE

Dark Fan-throated Lizard *Sitana fusca* NE

Broad-headed Fan-throated Lizard *Sitana laticeps* NE

Shore Fan-throated Lizard *Sitana marudhamneydhal* NE

Eastern Fan-throated Lizard *Sitana ponticeriana* LC

Schleich's Fan-throated Lizard *Sitana schleichi* NE

Sivalik Fan-throated Lizard *Sitana sivalensis* NE

Spiny-headed Fan-throated Lizard *Sitana spinaecephalus* NE

Visir's Fan-throated Lizard *Sitana visiri* NE

Brilliant Ground Agama *Trapelus agilis* NE

Occelated Ground Agama *Trapelus megalonyx* NE

Red-throated Ground Agama *Trapelus rubrigularis* NE

Baluch Ground Agama *Trapelus ruderatus* LC

Chamaeleonidae (Chameleons)

Indian Chamaeleon *Chamaeleo zeylanicus* LC

Gekkonidae (True Geckos)

Iranian Spider Gecko *Agamura persica* NE

Batur Rock Gecko *Altiphylax baturensis* NE

Minton's Bent-toed Gecko *Altiphylax mintoni* NE

Stoliczka's Bent-toed Gecko *Altiphylax stoliczkai* NE

Baluch Rock Gecko *Bunopus tuberculatus* LC

Indian Golden Gecko *Calodactylodes aureus* LC

Sri Lankan Golden Gecko *Calodactylodes illingworthorum* NE

Adi's Day Gecko *Cnemaspis adii* NE

Alwis' Day Gecko *Cnemaspis alwisi* NE

Amith's Day Gecko *Cnemaspis amith* NE

Anderson's Day Gecko *Cnemaspis andersonii* NE

Assamese Day Gecko *Cnemaspis assamensis* NE

Southern Day Gecko *Cnemaspis australia* DD

Beddome's Day Gecko *Cnemaspis beddomei* DD

Boei's Day Gecko *Cnemaspis boei* NE

Highland Day Gecko *Cnemaspis clivicola* NE

Gemunu's Day Gecko *Cnemaspis gemunu* NE

Giri's Day Gecko *Cnemaspis girii* NE

Goanese Day Gecko *Cnemaspis goaensis* EN

Slender Day Gecko *Cnemaspis gracilis* LC

Bauer's Day Gecko *Cnemaspis heteropholis* NT

Indian Day Gecko *Cnemaspis indica* VU

Das' Day Gecko *Cnemaspis indraneildasii* VU

Jerdon's Day Gecko *Cnemaspis jerdonii* VU

Sri Lankan Ornate Day Gecko *Cnemaspis kallima* NE

Kandy Day Gecko *Cnemaspis kandiana* LC

Kolhapur Day Gecko *Cnemaspis kolhapurensis* DD

Kottiyoor Day Gecko *Cnemaspis kottiyoorensis* NE

Kumarasingh's Day Gecko *Cnemaspis kumarasinghei* NE

Flowery Day Gecko *Cnemaspis latha* NE

Coastal Day Gecko *Cnemaspis littoralis* DD

Jewel Day Gecko *Cnemaspis menikay* NE

Molligoda's Day Gecko *Cnemaspis molligodai* NE

Montane Day Gecko *Cnemaspis monticola* DD

Mysore Day Gecko *Cnemaspis mysoriensis* LC

Ponmudi Day Gecko *Cnemaspis nairi* NT

Nilgiri Day Gecko *Cnemaspis nilagirica* DD

Ornate Day Gecko *Cnemaspis ornata* NT

Ota's Day Gecko *Cnemaspis otai* VU

Little Day Gecko *Cnemaspis pava* NE

Phillips' Day Gecko *Cnemaspis phillipsi* NE

Small Day Gecko *Cnemaspis podihuna* LC

Paradise Day Gecko *Cnemaspis pulchra* NE

Spotted Day Gecko *Cnemaspis punctata* NE

Rajakaruna's Day Gecko *Cnemaspis rajakarunai* NE

Rammala Day Gecko *Cnemaspis rammalensis* NE

Ritigala Day Gecko *Cnemaspis ritigalensis* NE

Samanala Day Gecko *Cnemaspis samanalensis* NE

Ferguson's Day Gecko *Cnemaspis scalpensis* NE

Forest Day Gecko *Cnemaspis silvula* NE

Sispara Ghat Day Gecko *Cnemaspis sisparensis* NT

Rough-bellied Day Gecko *Cnemaspis tropidogaster* DD

Upendra's Day Gecko *Cnemaspis upendrai* NE

Captain Wick's Day Gecko *Cnemaspis wicksi* NE

Wynaad Day Gecko *Cnemaspis wynadensis* EN

Yercaud Day Gecko *Cnemaspis yercaudensis* LC

Comb-toed Gecko *Crossobamon eversmanni* NE

Sindh Sand Gecko *Crossobamon orientalis* NE

Nicobar Bent-toed Gecko *Cyrtodactylus adleri* LC

Ayeyarwady Bent-toed Gecko *Cyrtodactylus ayeyarwadyensis* DD

Chitral Rock Gecko *Cyrtodactylus chitralensis* NE

Kollegal Ground Gecko *Cyrtodactylus collegalensis* NE

Sinharaja Bent-toed Gecko *Cyrtodactylus cracens* NE

Deccan Ground Gecko *Cyrtodactylus deccanensis* LC

Edward Taylor's Bent-toed Gecko *Cyrtodactylus edwardtaylori* NE

Banded Bent-toed Gecko *Cyrtodactylus fasciolatus* NE

Bridled Bent-toed Gecko *Cyrtodactylus fraenatus* NE

Sikkimese Bent-toed Gecko *Cyrtodactylus gubernatoris* NT

Himalayan Bent-toed Gecko *Cyrtodactylus himalayanus* NE

Jeypore Bent-toed Gecko *Cyrtodactylus jeyporensis* NE

Khasi Hills Bent-toed Gecko *Cyrtodactylus khasiensis* NE

Lawder's Bent-toed Gecko *Cyrtodactylus lawderanus* NE

Smith's Bent-toed Gecko *Cyrtodactylus malcolmsmithi* DD

Comba's Bent-toed Gecko *Cyrtodactylus marcuscombaii* NE

Stoll's Bent-toed Gecko *Cyrtodactylus martinstollii* NE

Clouded Ground Gecko *Cyrtodactylus nebulosus* LC

Nepalese Bent-toed Gecko *Cyrtodactylus nepalensis* NE

Ramboda Bent-toed Gecko *Cyrtodactylus ramboda* NE

Rishi Valley Bent-toed Gecko *Cyrtodactylus rishivalleyensis* NE

Andaman Bent-toed Gecko *Cyrtodactylus rubidus* NE

Sri Lankan Bent-toed Gecko *Cyrtodactylus soba* NE

Forest Spotted Gecko *Cyrtodactylus speciosus* NE

Srilekha's Bent-toed Gecko *Cyrtodactylus srilekhae* NE

Ratwana Bent-toed Gecko *Cyrtodactylus subsolanus* NE

Spotted Ground Gecko *Cyrtodactylus triedrus* NT

Giri's Bent-toed Gecko *Cyrtodactylus varadgirii* NE

Blotched Ground Gecko *Cyrtodactylus yakhuna* NE

Wall's Bent-toed Gecko *Cyrtodactylus walli* NE

Makran Spider Gecko *Cyrtopodion agamuroides* LC

Aravalli Rock Gecko *Cyrtopodion aravallensis* NE

Baig's Rock Gecko *Cyrtopodion baigii* NE

Battal Rock Gecko *Cyrtopodion battalensis* NE

Baluch Rock Gecko *Cyrtopodion belaense* NE

Datta Rock Gecko *Cyrtopodion dattanense* NE

Fort Munroe Rock Gecko *Cyrtopodion fortmunroi* LC

Indus Rock Gecko *Cyrtopodion indusoani* NE

Warty Rock Gecko *Cyrtopodion kachhense* NE

Koh Sulaiman Range Rock Gecko *Cyrtopodion kohsulaimanai* LC

Jammu Bent-toed Gecko *Cyrtopodion mansarulus* NE

Salt Range Thin-toed Gecko *Cyrtopodion montiumsalsorum* NE

Nepalese Rock Gecko *Cyrtopodion nepalensis* NE

Potohar Rock Gecko *Cyrtopodion potoharensis* NE

Red-tailed Rock Gecko *Cyrtopodion rhodocauda*

Rohtas Fort Rock Gecko *Cyrtopodion rohtasfortai* NE

Keeled Rock Gecko *Cyrtopodion scabrum* LC

Watson's Rock Gecko *Cyrtopodion watsoni* NE

Four-clawed Gecko *Gehyra mutilata* NE

Tokay Gecko *Gekko gecko* NE

Smith's Giant Gecko *Gekko smithii* LC

Andaman's Giant Gecko *Gekko verreauxi* NE

Bauer's Gecko *Hemidactylus aaronbaueri* LC

Tirunelveli Gecko *Hemidactylus acanthopholis* NE

White-banded Gecko *Hemidactylus albofasciatus* VU

Anaimalai Gecko *Hemidactylus anamallensis* NT

False Bowring's Gecko *Hemidactylus aquilonius* NE

Brooke's House Gecko *Hemidactylus brookii* NE

Depressed Gecko *Hemidactylus depressus* LC

Yellow-green House Gecko *Hemidactylus flaviviridis* NE

Asian House Gecko *Hemidactylus frenatus* LC

Garnot's Gecko *Hemidactylus garnotii* NE

Giant Southern Tree Gecko *Hemidactylus giganteus* LC

Gleadow's House Gecko *Hemidactylus gleadowi* NE

Slender Leaf-toed Gecko *Hemidactylus gracilis* LC

Granite Gecko *Hemidactylus graniticolus* LC

Gujarat Gecko *Hemidactylus gujaratensis* VU

Hemchandra's Gecko *Hemidactylus hemchandrai* NE

Sri Lankan Spotted Rock Gecko *Hemidactylus hunae* NE

Carrot-tailed Viper Gecko *Hemidactylus imbricatus* LC

Kushmore House Gecko *Hemidactylus kushmorensis* NE

Sri Lankan Termite Hill Gecko *Hemidactylus lankae* NE

Bark Gecko *Hemidactylus leschenaultii* NE

Spotted Rock Gecko *Hemidactylus maculatus* LC

Mahendra's Gecko *Hemidactylus mahendrai* NE

Murray's House Gecko *Hemidactylus murrayi* NE

Sri Lankan Spotted Gecko *Hemidactylus parvimaculatus* NE

Persian House Gecko *Hemidactylus persicus* NE

Piers' Gecko *Hemidactylus pieresii* NE

Flat-tailed Gecko *Hemidactylus platyurus* NE

Prashad's House Gecko *Hemidactylus prashadi* LC

Reticulated Gecko *Hemidactylus reticulatus* LC

Hayden's Gecko *Hemidactylus robustus* DD (as *Hemidactylus porbandarensis*)

Satara Gecko *Hemidactylus sataraensis* VU

Scaly Gecko *Hemidactylus scabriceps* DD

Treutler's House Gecko *Hemidactylus treutleri* LC

Termite-hill Gecko *Hemidactylus triedrus* NE

Turkish House Gecko *Hemidactylus turcicus* LC

Kanker Rock Gecko *Hemidactylus yajurvedi* NE

Western Ghats Worm Gecko *Hemiphyllodactylus aurantiacus* LC

Oceanic Worm Gecko *Hemiphyllodactylus typus* NE

Mourning Gecko *Lepidodactylus lugubris* NE

Short-limbed Gecko *Mediodactylus brachykolon* NE

Sindh Short-limbed Gecko *Mediodactylus sagittifer* NE

Wall's Short-limbed Gecko *Mediodactylus walli* NE

Baluch Dwarf Gecko *Microgecko depressus* LC

Khuzestan Dwarf Gecko *Microgecko helenae* DD

Persian Sand Gecko *Microgecko persicus* NE

Andaman Day Gecko *Phelsuma andamanense* LC

Blanford's Semaphore Gecko *Pristurus rupestris* LC

Burmese Gliding Gecko *Ptychozoon lionotum* LC

Nicobarese Gliding Gecko *Ptychozoon nicobarensis* NE

Pakistan Fan-fingered Gecko *Ptyodactylus homolepis* NE

Horned Gecko *Rhinogekko femoralis* NE

Missone's Spider Gecko *Rhinogekko misonnei* NE

Fedtschenko's Rock Gecko *Tenuidactylus fedtschenkoi* NE

Small-scaled Wonder Gecko *Teratoscincus microlepis* NE

Common Wonder Gecko *Teratoscincus scincus* NE

Eublepharidae (Eyelid Geckos)

Western Indian Leopard Gecko *Eublepharis fuscus* LC

Eastern Indian Leopard Gecko *Eublepharis hardwickii* LC

Commn Asian Leopard Gecko *Eublepharis macularius* NE

Satpura Leopard Gecko *Eublepharis satpuraensis* NE

Lacertidae (Lacertas)

Mekran Fringe-toed Lizard *Acanthodactylus blanfordii* NE

Indian Fringe-toed Lizard *Acanthodactylus cantoris* NE

Yellow-tailed Fringe-toed Lizard *Acanthodactylus micropholis* NE

Sharp-nosed Racerunner *Eremias acutirostris* LC

Cholistan Racerunner *Eremias cholistanica* NE

Banded Racerunner *Eremias fasciata* NE

Aralo-Caspian Racerunner *Eremias persica* NE

Sand Racerunner *Eremias scripta* NE

Short-nosed Lacerta *Mesalina brevirostris* LC

Desert Lacerta *Mesalina guttulata* NE

Watson's Lacerta *Mesalina watsonana* NE

Beddome's Lacerta *Ophisops beddomei* LC

Elegant Lacerta *Ophisops elegans* NE

Snake-eyed Lacerta *Ophisops jerdoni* LC

Leschenaulti's Lacerta *Ophisops leschenaultii* NE

Small-scaled Lacerta *Ophisops microlepis* LC

Small Lacerta *Ophisops minor* NE

Haughton's Long-tailed Lizard *Takydromus haughtonianus* NE

Khasi Hills Long-tailed Lizard *Takydromus khasiensis* NE

Sikkim Long-tailed Lizard *Takydromus sikkimensis* NE

Scincidae (Skinks)

Small Snake-eyed Skink *Ablepharus grayanus* NE

Asian Snake-eyed Skink *Ablepharus pannonicus* NE

Himalayan Rock Skink *Asymblepharus himalayanus* NE

Ladakhi Rock Skink *Asymblepharus ladacensis* NE

Mahabharat Range Rock Skink *Asymblepharus*

mahabharatus NE

Nepalese Rock Skink *Asymblepharus nepalensis* NE

Sikkimese Rock Skink *Asymblepharus sikimmensis* NE

Tragbul Pass Rock Skink *Asymblepharus tragbulensis* NE

Barkud Legless Skink *Barkudia insularis* CR

Vizag Legless Skink *Barkudia melanosticta* DD

Occelated Skink *Chalcides ocellatus* NE

Five-toed Skink *Chalcides pentadactylus* DD

Four-toed Skink *Chalcidoseps thwaitesii* NE

Haly's Tree Skink *Dasia halianus* NE

Johnsingh's Tree Skink *Dasia johnsinghi* NE

Nicobarese Tree Skink *Dasia nicobarensis* NE

Olive Tree Skink *Dasia olivacea* LC

Blue-bellied Tree Skink *Dasia subcaerulea* EN

Afridi Mole Skink *Eumeces blythianus* NE

Cholistan Mole Skink *Eumeces cholistanensis* NE

Indus Mole Skink *Eumeces indothalensis* NE

Pune Mole Skink *Eurylepis poonaensis* EN

Orange-tailed Mole Skink *Eumeces schneiderii* NE

Yellow-bellied Mole Skink *Eurylepis taeniolatus* NE

Allapal Grass Skink *Eutropis allapallensis* LC

Andaman Grass Skink *Eutropis andamanensis* NE

Austin's Grass Skink *Eutropis austini* NE

Beddome's Grass Skink *Eutropis beddomei* NE

Bibron's Sand Skink *Eutropis bibronii* LC

Keeled Grass Skink *Eutropis carinata* LC

Inger's Grass Skink *Eutropis clivicola* EN

Striped Grass Skink *Eutropis dissimilis* NE

Flower's Grass Skink *Eutropis floweri* NE

Gans' Grass Skink *Eutropis gansi* DD

Greer's Grass Skink *Eutropis greeri* NE

Blanford's Grass Skink *Eutropis innotata* DD

Bronze Grass Skink *Eutropis macularia* NE

Spotted Grass Skink *Eutropis madaraszi* NE

Many-lined Grass Skink *Eutropis multifasciata* NE

Nagarjun Grass Skink *Eutropis nagarjuni* NT

Nine-keeled Grass Skink *Eutropis novemcarinata* LC

Four-lobed Grass Skink *Eutropis quadratiloba* NE

Four-keeled Grass Skink *Eutropis quadricarinata* NE

Lined Grass Skink *Eutropis rudis* NE

Rough-backed Grass Skink *Eutropis rugifera* NE

Tammenna Skink *Eutropis tamanna* NE

Three-striped Skink *Eutropis trivittata* LC

Tytler's Grass Skink *Eutropis tytlerii* NE

Golden Grass Skink *Heremites auratus* NE

Beddome's Ground Skink *Kaestlea beddomii* LC

Two-lined Groung Skink *Kaestlea bilineatum* LC

Side-spotted Ground Skink *Kaestlea laterimaculata* VU

Palni Hills Ground Skink *Kaestlea palnica* DD

Travancore Ground Skink *Kaestlea travancorica* LC

Deignan's Lanka Skink *Lankascincus deignani* EN

Deraniyagala's Lanka Skink *Lankascincus deraniyagalae* NE

Chained Lanka Skink *Lankascincus dorsicatenatus* NE

Common Lanka Skink *Lankascincus fallax* NE

Gans's Lanka Skink *Lankascincus gansi* NE

Greer's Lanka Skink *Lankascincus greeri* NE

Munindradasa's Lanka Skink *Lankascincus munindradasai* NE

Sripada Lanka Skink *Lankascincus sripadensis* NE

Smooth Lanka Skink *Lankascincus taprobanensis* NT

Taylor's Lanka Skink *Lankascincus taylori* NE

Small-eared Striped Skink *Lipinia macrotympana* NE

White-spotted Supple Skink *Lygosoma albopunctata* NE

Ashwamedh Supple Skink *Lygosoma ashwamedhi* VU

Bowring's Supple Skink *Lygosoma bowringii* NE

Goan Supple Skink *Lygosoma goaense* DD

Günther's Supple Skink *Lygosoma guentheri* LC

Lined Supple Skink *Lygosoma lineata* LC

Lineated Supple Skink *Lygosoma lineolatum* NE

Pruth's Supple Skink *Lygosoma pruthi* DD

Spotted Supple Skink *Lygosoma punctata* NE

Sri Lankan Supple Skink *Lygosoma singha* DD

Vosmaer's Supple Skink *Lygosoma vosmaerii* DD

Smith's Snake Skink *Nessia bipes* NE

Burton's Snake Skink *Nessia burtonii* NE

Deraniyagala's Snake Skink *Nessia deraniyagalai* NE

Two-toed Snake Skink *Nessia didactylus* NE

Sri Lankan Snake Skink *Nessia hickanala* NE

Layard's Snake Skink *Nessia layardi* NE

Toeless Snake Skink *Nessia monodactylus* NE

Sarasins' Snake Skink *Nessia sarasinorum* NE

Indian Sandfish *Ophiomorus raithmai* LC

Afghan Sandfish *Ophiomorus tridactylus* NE

Beddome's Cat Skink *Ristella beddomii* LC

Günther's Cat Skink *Ristella guentheri* DD

Rurk's Cat Skink *Ristella rurkii* DD

Travancore Cat Skink *Ristella travancoricus* DD

Large Ground Skink *Scincella capitanea* NE

Large-eared Ground Skink *Scincella macrotis* NE

Modest Ground Skink *Scincella modestum* NE

Reeves's Ground Skink *Scincella reevesii* NE

Spotted Snake Skink *Sepsophis punctatus* NE

Naked-eyed Litter Skink *Sphenomorphus apalpebrae* NE

Rotung Litter Skink *Sphenomorphus courcyanum* NE

Dussumier's Litter Skink *Sphenomorphus dussumieri* LC

Himalayan Litter Skink *Sphenomorphus indicus* NE

Spotted Litter Skink *Sphenomorphus maculatus* NE

North-eastern Water Skink *Tropidophorus assamensis* NE

Anguidae (Glass Snakes)

Eastern Glass Snake *Dopasia gracilis* NE

Dibamidae (Worm Lizards)

Nicobarese Worm Lizard *Dibamus nicobaricum* NE

Varanidae (Monitors)

Bengal Monitor *Varanus bengalensis* LC

Yellow Monitor *Varanus flavescens* LR/LC

Desert Monitor *Varanus griseus* NE

Water Monitor *Varanus salvator* LC

Acrochordidae (Wart Snakes)

Wart Snake *Acrochordus granulatus* LC

Cylindrophiidae (Pipe Snakes)

Sri Lankan Pipe Snake *Cylindrophis maculata* NE

Uropeltidae (Shieldtailed Snakes)

Palni Shieldtail *Brachyophidium rhodogaster* LC

Yellow-striped Shieldtail *Melanophidium bilineatum* VU

Khaire's Shieldtail *Melanophidium khairei* NE

Pied-bellied Shieldtail *Melanophidium punctatum* LC

Wynad Shieldtail *Melanophidium wynaudensis* LC

Madurai Shieldtail *Platyplectrurus madurensis* EN

Travancore Shieldtail *Platyplectrurus ruhunae* NE

Three-lined Shiledtail *Platyplectrurus trilineatus* DD

Kerala Shieldtail *Plectrurus aureus* DD

Kanara Shieldtail *Plectrurus canaricus* DD

Günther's Shieldtail *Plectrurus guentheri* DD

Nilgiri Shieldtail *Plectrurus perroteti* LC

Large Shieldtail *Pseudotyphlops philippinus* NE

Blyth's Shieldtail *Rhinophis blythii* NE

Polka-dot Shieldtail *Rhinophis dorsimaculatus* NE

Drummond-Hay's Shieldtail *Rhinophis drummondhayi* NT

Eranga Viraj's Shieldtail *Rhinophis erangaviraji* NE
Cardamom Hills Shieldtail *Rhinophis fergusonianus* DD
Gower's Shieldtail *Rhinophis goweri* NE
Hemprich's Shieldtail *Rhinophis homolepis* NE
Lined Shieldtail *Rhinophis lineatus* NE
Black-bellied Shieldtail *Rhinophis melanogaster* NE
Schneider's Shieldtail *Rhinophis oxyrynchus* LC
Cuvier's Shieldtail *Rhinophis philippinus* NE
Phillips' Shieldtail *Rhinophis phillipsi* NE
Giant Shieldtail *Rhinophis porrectus* NE
Spotted Shieldtail *Rhinophis punctatus* NE
Kelaart's Shieldtail *Rhinophis saffragamus* NE
Red-bellied Shieldtail *Rhinophis sanguineus* LC
Travancore Shieldtail *Rhinophis travancoricus* EN
Tricoloured Shieldtail *Rhinophis tricoloratus* NE
Zigzag Shieldtail *Rhinophis zigzag* NE
Purple-red Shieldtail *Teretrurus sanguineus* LC
Tirunelveli Shieldtail *Uropeltis arcticeps* LC
Beddome's Shieldtail *Uropeltis beddomii* DD
Twin-chained Shieldtail *Uropeltis bicatenata* NT
Brougham's Shieldtail *Uropeltis broughami* DD
Cuvier's Shieldtail *Uropeltis ceylanica* LC
Sirumalai Shieldtail *Uropeltis dindigalensis* DD
Elliot's Shieldtail *Uropeltis ellioti* LC
Ashambu Shieldtail *Uropeltis liura* DD
Large-scaled Shieldtail *Uropeltis macrolepis* LC
Anaimalai Shieldtail *Uropeltis macrorhyncha* DD
Spotted Shieldtail *Uropeltis maculata* DD
Boulenger's Shieldtail *Uropeltis myhendrae* DD
Shiny Shieldtail *Uropeltis nitida* DD
Occelated Shieldtail *Uropeltis ocellata* LC
Peters' Shieldtail *Uropeltis petersi* DD
Phillip's Shieldtail *Uropeltis phillipsi* NE
Phipson's Shieldtail *Uropeltis phipsonii* VU
Palni Hills Shieldtail *Uropeltis pulneyensis* LC
Red-lined Shieldtail *Uropeltis rubrolineata* LC
Red-spotted Shieldtail *Uropeltis rubromaculata* LC
Southern Shieldtail *Uropeltis ruhunae* NE
Shevaroy Shieldtail *Uropeltis shorttii* NE
Smith's Shieldtail *Uropeltis smithi* NT
Woodmason's Shieldtail *Uropeltis woodmasoni* LC

Pythonidae (Pythons)

Reticulated Python *Malayopython reticulatus* NE
Burmese Rock Python *Python bivittatus* VU
Indian Rock Python *Python molurus* NE

Xenopeltidae (Sunbeam Snakes)

Sunbeam Snake *Xenopeltis unicolor* LC

Boidae (Boas)

Common Sand Boa *Eryx conicus* NE
Red Sand Boa *Eryx johnii* NE
Tartary Sand Boa *Eryx tataricus* NE
Whitaker's Sand Boa *Eryx whitakeri* NE

Colubridae (Typical Snakes)

Günther's Vine Snake *Ahaetulla dispar* NT
Variable Coloured Vine Snake *Ahaetulla anomala*
Common Vine Snake *Ahaetulla nasuta* NE
Perrotet's Vine Snake *Ahaetulla perroteti* EN
Oriental Vine Snake *Ahaetulla prasina* LC
Brown Vine Snake *Ahaetulla pulverulenta* LC
Dice-like Trinket Snake *Archelaphe bella* LC
Banded Racer *Argyrogena fasciolata* NE
Stripe-tailed Racer *Argyrogena vittacaudata* NE
Mizo Iridescent Snake *Blythia hmuifang*
Iridescent Snake *Blythia reticulata* DD
Andamans Cat Snake *Boiga andamanensis* NE
Barnes' Cat Snake *Boiga barnesii* NE
Beddome's Cat Snake *Boiga beddomei* DD
Sri Lankan Cat Snake *Boiga ceylonensis* NE
Green Cat Snake *Boiga cyanea* NE
Travancore Cat Snake *Boiga dightoni* DD
Yellow-green Cat Snake *Boiga flaviviridis* NE
Forsten's Cat Snake *Boiga forsteni* LC
Eastern Cat Snake *Boiga gokool* NE

Many-banded Cat Snake *Boiga multifasciata* DD

Many-spotted Cat Snake *Boiga multomaculata* NE

Collared Cat Snake *Boiga nuchalis* NE

Tawny Cat Snake *Boiga ochracea* NE

Assamese Cat Snake *Boiga quincunciata* NE

Thai Cat Snake *Boiga siamensis* NE

Common Indian Cat Snake *Boiga trigonata* LC

Nicobarese Cat Snake *Boiga wallachi* DD

Reed Snake *Calamaria pavimentata* LC

Ornate Flying Snake *Chrysopelea ornata* NE

Red-spotted Flying Snake *Chrysopelea paradisi* LC

Sri Lankan Flying Snake *Chrysopelea taprobanica* NE

Yellow-striped Trinket Snake *Coelognathus flavolineatus* LC

Indian Trinket Snake *Coelognathus helena* NE

Copper-headed Trinket Snake *Coelognathus radiatus* LC

Indian Smooth Snake *Coronella brachyura* LC

Doria's Green Snake *Cyclophiops doriae* NE

Andaman Bronzeback Tree Snake *Dendrelaphis andamanensis* NE

Ashok's Bronzeback Tree Snake *Dendrelaphis ashoki* LC

Boulenger's Bronzeback Tree Snake *Dendrelaphis bifrenalis* LC

Wall's Bronzeback Tree Snake *Dendrelaphis biloreatus* LC (as *Dendrelaphis gorei*)

Stripe-tailed Bronzeback Tree Snake *Dendrelaphis caudolineolatus* NE

Boie's Bronzeback Tree Snake *Dendrelaphis chairecacos* DD

Blue Bronzeback Tree Snake *Dendrelaphis cyanochloris* LC

Giri's Bronzeback Tree Snake *Dendrelaphis girii* LC

Large-eyed Tree Snake *Dendrelaphis grandoculis* LC

Nicobar Bronzeback Tree Snake *Dendrelaphis humayuni* NE

Oliver's Bronzeback Tree Snake *Dendrelaphis oliveri* NE

Painted Bronzeback Tree Snake *Dendrelaphis proarchos* NE

Mountain Bronzeback Tree Snake *Dendrelaphis subocularis* LC

Kuhl's Bronzeback Tree Snake *Dendrelaphis schokari* NE

Sinharaja Bronzeback Tree Snake *Dendrelaphis sinharajensis* NE

Common Bronzeback Tree Snake *Dendrelaphis tristis* NE

Scarce Bridle Snake *Dryocalamus gracilis* DD

Vellore Bridle Snake *Dryocalamus nympha* NE

Dark-headed Dwarf Racer *Eirenis persicus* NE

Indian Egg-eating Snake *Elachistodon westermanni* LC

Mandarin Trinket Snake *Euprepiophis mandarinus* LC

Nicobarese Ring-necked Snake *Gongylosoma nicobariensis* NE

Khasi Hills Trinket Snake *Gonyosoma frenatum* NE

Red-tailed Trinket Snake *Gonyosoma oxycephalum* LC

Green Trinket Snake *Gonyosoma prasinum* LC

Spotted Whip Snake *Hemorrhois ravergieri* NE

Reed-like Stripe-necked Snake *Liopeltis calamaria* NE

Stripe-necked Snake *Liopeltis frenatus* LC

Himalayan Stripe-necked Snake *Liopeltis rappii* DD

Stoliczka's Stripe-necked Snake *Liopeltis stoliczkae* LC

Common Wolf Snake *Lycodon aulicus* NE

Sri Lanka Wolf Snake *Lycodon carinatus* NE

Banded Wolf Snake *Lycodon fasciatus* NE

Yellow-collared Wolf Snake *Lycodon flavicollis* NE

Yellow-spotted Wolf Snake *Lycodon flavomaculatus* LC

Sikkimese Wolf Snake *Lycodon gammiei* NE

Andamans Wolf Snake *Lycodon hypsirhinoides* NE

Yellow-speckled Wolf Snake *Lycodon jara* LC

Laotian Wolf Snake *Lycodon laoensis* LC

MacKinnon's Wolf Snake *Lycodon mackinnoni* NE

Flowered Wolf Snake *Lycodon osmanhilli* LC

Large-toothed Wolf Snake *Lycodon septentrionalis* NE

Barred Wolf Snake *Lycodon striatus* NE

Tiwari's Wolf Snake *Lycodon tiwarii* NE

Travancore Wolf Snake *Lycodon travancoricus* LC

Zaw's Wolf Snake *Lycodon zawi* LC

Maynard's Long-nosed Sand Snake *Lytorhynchus maynardi* LC

Sindh Long-nosed Sand Snake *Lytorhynchus paradoxus* NE

Afghan Awl-headed Snake *Lytorhynchus ridgewayi* LC

Günther's Kukri Snake *Oligodon affinis* LC

White-barred Kukri Snake *Oligodon albocinctus* NE

Banded Kukri Snake *Oligodon arnensis* NE

Short-tailed Kukri Snake *Oligodon brevicauda* VU

Reed-like Kukri Snake *Oligodon calamarius* NE

Assamese Kukri Snake *Oligodon catenatus* NE

Grey Kukri Snake *Oligodon cinereus* LC

Cantor's Kukri Snake *Oligodon cyclurus* LC

Spot-tailed Kukri Snake *Oligodon dorsalis* NE

Red-bellied Kukri Snake *Oligodon erythrogaster* NE

Namsang Kukri Snake *Oligodon erythrorhachis* DD

Wall's Kukri Snake *Oligodon juglandifer* VU

Kheri Kukri Snake *Oligodon kheriensis* NE

Blue-bellied Kukri Snake *Oligodon melaneus* NE

Abor Hills Kukri Snake *Oligodon melanozonatus* NE

Nikhil's Kukri Snake *Oligodon nikhili* DD

Duméril's Kukri Snake *Oligodon sublineatus* LC

Streaked Kukri Snake *Oligodon taeniolatus* LC

Mandalay Kukri Snake *Oligodon theobaldi* LC

Travancore Kukri Snake *Oligodon travancoricus* DD

Jerdon's Kukri Snake *Oligodon venustus* LC

Nicobarese Kukri Snake *Oligodon woodmasoni* NE

Black-banded Trinket Snake *Oreocryptophis porphyraceus* NE

Eastern Trinket Snake *Orthriophis cantoris* NE

Himalayan Trinket Snake *Orthriophis hodgsonii* NE

Striped Trinket Snake *Orthriophis taeniurus* NE

Bhola Nath's Racer *Platyceps bholanathi* DD

Slender Racer *Platyceps gracilis* DD

Spotted Desert Racer *Platyceps karelini* NE

Cliff Racer *Platyceps ladacensis* NE

Noel's Racer *Platyceps noeli* NE

Striped Racer *Platyceps rhodorachis* NE

Sindh Racer *Platyceps sindhensis* NE

Glossy-bellied Racer *Platyceps ventromaculatus* NE

Mock Viper *Psammodynastes pulverulentus* NE

Eastern Rat Snake *Ptyas korros* NE

Indian Rat Snake *Ptyas mucosa* NE

Green Rat Snake *Ptyas nigromarginata* NE

Two-coloured Forest Snake *Rhabdops bicolor* NE

Olive Forest Snake *Rhabdops olivaceus* LC

Two-lined Black-headed Snake *Sibynophis bistrigatus* DD

Collared Black-headed Snake *Sibynophis collaris* LC

Cantor's Black-headed Snake *Sibynophis sagittarius* NE

Black-headed Snake *Sibynophis subpunctatus* NE

Red-spotted Royal Snake *Spalerosophis arenarius* NE

Black-headed Royal Snake *Spalerosophis atriceps* NE

Diadem Snake *Spalerosophis diadema* NE

Desert Tiger Snake *Telescopus rhinopoma* LC

Darjeeling Slender Snake *Trachischium fuscum* NE

Günther's Slender Snake *Trachischium guentheri* LC

Olive Slender Snake *Trachischium laeve* NE

Assamese Slender Snake *Trachischium monticola* NE

Slender-headed Snake *Trachischium tenuiceps* NE

Wallace's Racer *Wallaceophis gujaratensis* NE

Lamprophiidae (Sand Snakes)

Oriental Sand Snake *Psammophis condanarus* LC

Pakistani Ribbon Snake *Psammophis leithii* NE

Steppe Ribbon Snake *Psammophis lineolatus* NE

Long Sand Snake *Psammophis longifrons* LC

Forskål's Sand Snake *Psammophis schokari* NE

Natricidae (Keelback Snakes)

Buff-striped Keelback *Amphiesma stolatum* NE

Boie's Rough-sided Snake *Aspidura brachyorrhos* NE

Black-spined Snake *Aspidura ceylonensis* NE

Cope's Rough-sided Snake *Aspidura copei* DD

Deraniyagala's Rough-sided Snake *Aspidura deraniyagalae* NE

Drummond-Hay's Rough-sided Snake *Aspidura drummondhayi* NE

Günther's Rough-sided Snake *Aspidura guentheri* NE

Common Rough-sided Snake *Aspidura trachyprocta* NE

Olive Keelback Water Snake *Atretium schistosum* LC

Beddome's Keelback *Hebius beddomii* LC

Clerk's Keelback *Hebius clerki* NE

Khasi Hills Keelback *Hebius khasiense* NE

Günther's Keelback *Hebius modestum* LC

Montane Keelback *Hebius monticola* LC

Nicobarese Keelback *Hebius nicobarense* NE

Boulenger's Keelback *Hebius parallelum* NE

Peal's Keelback *Hebius pealii* NE

Venning's Keelback *Hebius venningi* LC

Strange-tailed Keelback *Hebius xenura* NE

Eastern Keelback *Herpetoreas platyceps* NE

Siebold's Keelback *Herpetoreas sieboldii* DD

Green Keelback *Macropisthodon plumbicolor* NE

Dice Snake *Natrix tessellata* LC

Sri Lankan Keelback *Rhabdophis ceylonensis* NT

Himalayan Keelback *Rhabdophis himalayanus* NE

Collared Keelback *Rhabdophis nuchalis* LC

Red-necked Keelback *Rhabdophis subminiatus* LC

Eastern Keelback Water Snake *Sinonatrix percarinata* LC

Sri Lankan Keelback Water Snake *Xenochrophis asperrimus* NE

Dark-bellied Keelback Water Snake *Xenochrophis cerasogaster* NE

Yellow-spotted Keelback Water Snake *Xenochrophis flavipunctatus* LC

Checkered Keelback Water Snake *Xenochrophis piscator* NE

Spotted Keelback Water Snake *Xenochrophis punctulatus* LC

St. John's Keelback Water Snake *Xenochrophis sanctijohannis* NE

Schnurrenberger's Keelback Water Snake *Xenochrophis schnurrenbergeri* NE

Triangled Keelback Water Snake *Xenochropis trianguligerus* LC

Andamans Keelback Water Snake *Xenochrophis tytleri* NE

Pseudoxenodontidae (False Cobras)

Large-eyed False Cobra *Pseudoxenodon macrops* LC

Elapidae (Cobras, Kraits, Coral Snakes and Sea Snakes)

Andamans Krait *Bungarus andamanensis* VU

Himalayan Krait *Bungarus bungaroides* NE

Indian Krait *Bungarus caeruleus* NE

Sri Lankan Krait *Bungarus ceylonicus* NE

Banded Krait *Bungarus fasciatus* LC

Lesser Black Krait *Bungarus lividus* NE

Black Krait *Bungarus niger* NE

Sindh Krait *Bungarus sindanus* NE

Beddome's Coral Snake *Calliophis beddomei* DD

Bibron's Coral Snake *Calliophis bibroni* LC

Castoe's Coral Snake *Calliophis castoe* NE

Red-bellied Coral Snake *Calliophis haematoetron* NE

Indian Coral Snake *Calliophis melanurus* NE

Black Slender Coral Snake *Calliophis*

nigrescens LC

Peters' Sea Snake *Hydrophis bituberculatus* DD

Swarf Sea Snake *Hydrophis caerulescens* LC

Cantor's Narrow-headed Sea Snake *Hydrophis cantoris* DD

Short Sea Snake *Hydrophis curtus* LC

Annulated Sea Snake *Hydrophis cyanocinctus* LC

Banded Sea Snake *Hydrophis fasciatus* LC

Slender Sea Snake *Hydrophis gracilis* LC

Jerdon's Sea Snake *Hydrophis jerdonii* LC

Persian Gulf Sea Snake *Hydrophis lapemoides* LC

Bombay Sea Snake *Hydrophis mamillaris* DD

Black-banded Sea Snake *Hydrophis nigrocinctus* DD

Russell's Sea Snake *Hydrophis obscura* LC

Ornate Sea Snake *Hydrophis ornatus* LC

Pelagic Sea Snake *Hydrophis platurus* LC

Hook-nosed Sea Snake *Hydrophis schistosus* LC

Yellow Sea Snake *Hydrophis spiralis* LC

Stokes' Sea Snake *Hydrophis stokesii* LC

Collared Sea Snake *Hydrophis stricticollis* DD

Olive Sea Snake *Hydrophis viperinus* LC

Yellow-lipped Sea Krait *Laticauda colubrina* LC

Blue-lipped Sea Krait *Laticauda laticaudata* LC

Monocled Cobra *Naja kaouthia* LC

Spectacled Cobra *Naja naja* NE

Central Asian Cobra *Naja oxiana* DD

Andaman Cobra *Naja sagittifera* NE

King Cobra *Ophiophagus hannah* VU

MacClelland's Coral Snake *Sinomicrurus macclellandi* NE

Homalopsidae (Mud Snakes)

Yellow-banded Mangrove Snake *Cantoria violacea* LC

Dog-faced Water Snake *Cerberus rynchops* LC

Dussumier's Water Snake *Enhydris dussumierii* LC

Common Smooth Water Snake *Enhydris enhydris* LC

Siebold's Water Snake *Enhydris sieboldii* LC

White-bellied Mangrove Snake *Fordonia leucobalia* LC

Glossy Marsh Snake *Gerarda prevostiana* LC

Puff-faced Water Snake *Homalopsis buccata* LC

Hardwick's Water Snake *Homalopsis hardwickii* NE

Pakistani Water Snake *Mintonophis pakistanicus* LC

Pareidae (Slug-eating Snakes)

Spotted Slug-eating Snake *Pareas macularius* NE

Montane Slug-eating Snake *Pareas monticolus* NE

Viperidae (Vipers and Pit Vipers)

Malabar Pit Viper *Craspedocephalus malabaricus* LC

Horseshoe Pit Viper *Craspedocephalus strigatus* LC

White-lipped Pit Viper *Cryptelytrops albolabris* LC

Anderson's Pit Viper *Cryptelytrops andersoni* NE

Cantor's Pit Viper *Cryptelytrops cantori* NE

Red-tailed Pit Viper *Cryptelytrops erythrurus* LC

Nicobar Pit Viper *Cryptelytrops labialis* NE

Central Nicobar Pit Viper *Cryptelytrops mutabilis* NE

Spot-tailed Pit Viper *Cryptelytrops septentrionalis* NE

Russell's Viper *Daboia russelii* NE

Saw-scaled Viper *Echis carinatus* NE

Leaf-nosed Viper *Eristicophis macmahoni* NE

Himalayan Pit Viper *Gloydius himalayanus* NE

Tibetan Pit Viper *Himalayophis tibetanus* LC

Hump-nosed Pit Viper *Hypnale hypnale* NE

Millard's Hump-nosed Pit Viper *Hypnale nepa* LC

Gray's Hump-nosed Pit Viper *Hypnale zara* NE

Levantine Viper *Macrovipera lebetina* NE

Blotched Pit Viper *Ovophis monticola* LC

Large-scaled Pit Viper *Peltopelor macrolepis* NT

Pope's Pit Viper *Popeia popeiorum* LC

Eastern Himalayan Pit Viper *Protobothrops*

himalayanus NE

Jerdon's Pit Viper *Protobothrops jerdonii* LC

Kaulback's Pit Viper *Protobothrops kaulbacki* DD

Brown-spotted Pit Viper *Protobothrops mucrosquamatus* LC

Persian Horned Viper *Pseudocerastes persica* LC

Bamboo Pit Viper *Trimeresurus gramineus* LC

Sri Lankan Green Pit Viper *Trimeresurus trigonocephalus* NE

Hutton's Pit Viper *Tropidolaemus huttoni* NE

Gumprecht's Pit Viper *Viridovipera gumprechti* LC

Medo Pit Viper *Viridovipera medoensis* DD

Stejneger's Pit Viper *Viridovipera stejnegeri* LC

Yunnan Pit Viper *Viridovipera yunnanensis* LC

Xenodermatidae (Xenodermatid Snakes)

Khasi Hills Snake *Stoliczkia khasiensis* NE

Captain's Wood Snake *Xylophis captaini* LC

Perrotet's Wood Snake *Xylophis perroteti* LC

Günther's Wood Snake *Xylophis stenorhynchus* DD

Gerrhopilidae (South Asian Blind Snakes)

Andamans Blind Snake *Gerrhopilus andamanensis* NE

Beddome's Blind Snake *Gerrhopilus beddomei* DD

Sri Lankan Blind Snake *Gerrhopilus ceylonicus* NE

Jan's Blind Snake *Gerrhopilus mirus* NE

Wall's Blind Snake *Gerrhopilus oligolepis* DD

Thurston's Blind Snake *Gerrhopilus thurstoni* NE

Tindall's Blind Snake *Gerrhopilus tindalli* DD

Typhlopidae (Blind Snakes)

Günther's Blind Snake *Argyrophis bothriorhynchus* DD

Large Blind Snake *Argyrophis diardii* LC

Narrow-necked Blind Snake *Asiatyphlops tenuicollis* DD

Beaked Blind Snake *Grypotyphlops acutus* LC

Brahminy Blind Snake *Indotyphlops braminus* NE

Belgaum Blind Snake *Indotyphlops exiguus* DD

Jerdon's Blind Snake *Indotyphlops jerdoni* NE

Sri Lankan Blind Snake *Indotyphlops lankaensis* NE

Pied Blind Snake *Indotyphlops leucomelas* NE

Loveridge's Blind Snake *Indotyphlops loveridgei* NE

Malcolm Smith's Blind Snake *Indotyphlops malcolmi* NE

Darjeeling Blind Snake *Indotyphlops meszoelyi* DD

Müller's Blind Snake *Indotyphlops muelleri* LC

Günther's Blind Snake *Indotyphlops pammeces* LC

Slender Blind Snake *Indotyphlops porrectus* NE

Taylor's Blind Snake *Indotyphlops tenebrarum* NE

Chumakedima Blind Snake *Indotyphlops tenuicollis* DD

Vedda Blind Snake *Indotyphlops veddae* NE

Violet Blind Snake *Indotyphlops violaceus* NE

Leptotyphlopidae (Thread Snakes)

Blanford's Thread Snake *Myriopholis blanfordii* NE

Large-beaked Thread Snake *Myriopholis macrorhyncha* NE

ORDER CROCODYLIA

Crocodylidae (Crocodiles)

Mugger Crocodile *Crocodylus palustris* VU

Saltwater Crocodile *Crocodylus porosus* LR/NT

Gavialiidae (Gharials)

Indian Gharial *Gavialis gangeticus* CR

Checklist current: 10 July 2017.

Further Reading

Daniel, J. C. 1983. *The Book of Indian Reptiles*. Oxford University Press/Bombay Natural History Society, Bombay. 141pp.

Das, I. 1996. *Biogeography of the Reptiles of South Asia*. Krieger Publishing Company, Malabar, Florida. 16 pl. + vii + 87pp.

Das, I. 2002. *A Photographic Guide to the Snakes and Other Reptiles of India*. New Holland Publishers (UK) Ltd, London. 144pp.

Das, I. 2010. *A Field Guide to the Reptiles of Southeast Asia*. New Holland Publishers (UK) Ltd., London. 376pp.

Das, I. & A. de Silva. 2005. *A Photographic Guide to the Snakes and Other Reptiles of Sri Lanka*. New Holland Publishers (UK) Ltd., London. 144pp.

Minton, S. A. 1966. A contribution to the herpetology of West Pakistan. *Bulletin of the American Museum of Natural History* 134:27–184.

Pyron, R. A., S. R. Ganesh, A. Sayyed, V. Sharma, V. Wallach & R. Somaweera. 2016. A catalogue and systematic overview of the shield-tailed snakes (Serpentes: Uropeltidae). *Zoosystema* 38(4):453–506.

Smith, M. A. 1931. *The Fauna of British India, Including Ceylon and Burma. Vol. I. Loricata, Testudines*. Taylor and Francis, London. xxviii + 185pp + 2pl.

Smith, M. A. 1935. *The Fauna of British India, Including Ceylon and Burma. Reptilia and Amphibia. Vol. II. Sauria*. Taylor and Francis, London. xiii + 440pp + 1pl.

Smith, M. A. 1943. *The Fauna of British India, Ceylon and Burma, Including the Whole of the Indo-Chinese Region. Vol. III. Serpentes*. Taylor and Francis, London. xii + 583pp. + 1 map.

Whitaker, R. 1978. *Common Indian Snakes: A Field Guide*. Macmillan India Limited, New Delhi. xiv + 154pp.

Whitaker, R. & A. S. Captain. 2004. *Snakes of India. The Field Fuide*. Draco Books, Chennai. xiv + 481pp.

Acknowledgements

We thank John Beaufoy for inviting us to write this book. Manuscript preparation was supported by Universiti Malaysia Sarawak and the Wildlife Institute of India, for which we thank our respective directors, Gabriel Tonga Noweg and Vinod Mathur.

For sharing papers and unpublished information, we thank the following colleagues: Kraig Adler, Ishan Agarwal, Farid Ahsan, Natalia Ananjeva, E. Nicholas Arnold, Kurt Auffenberg, the late Walter Auffenberg, Mark Auliya, Christopher Austin, David Barker, Aaron M. Bauer, the late Subramanian Bhupathy, Wolfgang Böhme, Rafe Brown, Ashok Captain, John Cadle, Lawan Chanhome, Binod Choudhury, Jennifer Daltry, the late Ilya Darevsky, Patrick David, Anslem de Silva, Kaushik Deuti, Sushil K. Dutta, Maren Gaulke, David Gower, Allen Greer, Wolfgang Grossmann, Andreas Gumprecht, Jakob Hallermann, Harold Heatwole, Ivan Ineich, Robert F. Inger, Jiang Jian-Ping, Ulrich Joger, Mohammad Sharief Khan, the late Vladimir Kharin, Alan Leviton, Kelvin K. P. Lim, Aaron Lobo, Colin McCarthy, Pratyush Mohapatra, Steven Mahony, Anita Malhotra, Edmond Malnate, Ulrich Manthey, the late Sherman Minton, Viral Mistry, John Murphy, B.H.C.K. Murthy, Nikolai Orlov, Mark O'Shea, Hidetoshi Ota, Steve Platt, Sujoy Raha, Arne Rasmussen, the late Jens Rasmussen, the late Charles A. Ross, Saibal Sengupta, Klaus-Dieter Schulz, Shailendra Singh, the late Joseph Slowinski, the late Hobart Smith, Frank Tillack, the late Garth Underwood, Karthik Vasudevan, Gernot Vogel, Harold Voris, Van Wallach, Romulus Whitaker, Guin Wogan, Wolfgang Wüster, Thomas Ziegler, Er-Mi Zhao and George Zug.

We are grateful to a number of agencies for research permits over the years from India and Sri Lanka. Field work was supported by grants from the Centre for Herpetology, Madras Crocodile Bank Trust, Wildlife Institute of India, World Wide Fund for Nature, World Conservation Union (IUCN), Convention on International Trade in Endangered Species of Wild Fauna and Flora, Center for Marine Conservation, Darwin Initiative for the Conservation of Species, Fauna and Flora International, People's Trust for Endangered Species, Aaranyak, Rufford Foundation UK and World Nature Association.

Additional images were contributed by Steven C. Anderson, Rajeev Basumatary, Ashok Captain, David Jones, Muhammad Sharief Khan, H. T. Lalremsanga, Manoj Nair, Nikolai Orlov, Klaus-Dieter Schulz, Alexander Teynié, Gernot Vogel, Raju Vyas and George R. Zug.

Finally, we thank Genevieve V. A. Gee for reading a draft of this work, Krystyna Mayer for her editorial work, and Rosemary Wilkinson for seeing the volume through press.

Index

Acanthodactylus cantoris 64
Acrochordus granulatus 83
Afghan Awl-headed Snake 111
Ahaetulla nasuta 91
   prasina 91
Amphiesma stolatum 122
Andaman Bent-toed Gecko 51
Andaman Bronzeback Tree Snake 100
Andaman Cobra 138
Andaman Day Gecko 62
Andaman Grass Skink 68
Andamans Keelback Water Snake 132
Andamans Wolf Snake 108
Anderson's Mountain Lizard 37
Anderson's Pit Viper 144
Annulated Sea Snake 136
Arakan Hill Turtle 20
Archelape bella 92
Argyrophis diardii 153
Ashambu Shieldtail 85
Asian Giant Softshell Turtle 26
Asian Giant Tortoise 14
Asian House Gecko 56
Aspidura ceylonensis 122
   guentheri 123
   trachyprocta 123
Assam Roofed Turtle 22
Assamese Cat Snake 96
Assamese Day Gecko 46
Assamese Slender Snake 121
Asymblepharus ladacensis 67
Atretium schistosum 124
Bamboo Pit Viper 151
Banded Bent-toed Gecko 50
Banded Krait 134
Banded Kukri Snake 112
Banded Wolf Snake 107
Bark Gecko 58
Barred Wolf Snake 110
Batagur baska 16
   dhongoka 16
   kachuga 17
Bay Islands Forest Lizard 36
Beddome's Keelback 124
Bengal Monitor 81

Black Krait 135
Black Softshell Turtle 25
Black-banded Trinket Snake 115
Black-headed Snake 120
Black-lipped Lizard 33
Black-spined Snake 122
Blotched Ground Gecko 52
Blotched Pit Viper 149
Blue Bronzeback Tree Snake 101
Blyth's Shieldtail 84
Blythia reticulata 92
Boiga ceylonensis 93
   cyanea 93
   forsteni 94
   gokool 94
   multomaculata 95
   ochracea 95
   quincunciata 96
   siamensis 96
   trigonata 97
   wallachi 97
Boulenger's Keelback 126
Bowring's Supple Skink 76
Brahminy Blind Snake 154
Brilliant Ground Agama 43
Bronze Grass Skink 70
Brooke's House Gecko 55
Brown Roofed Turtle 21
Brown-spotted Pit Viper 151
Buff-striped Keelback 122
Bufoniceps laungwalansis 29
Bungarus caeruleus 133
   ceylonicus 134
   fasciatus 134
   niger 135
Burmese Rock Python 87
Calliophis melanurus 135
Calodactylodes aureus 45
Calodactyloides illingworthorum 46
Calotes calotes 29
   ceylonensis 30
   emma 30
   jerdoni 31
   liolepis 31
   maria 32
   mystaceus 32
   nigrilabris 33

   versicolor 33
Cantor's Black-headed Snake 120
Cantor's Kukri Snake 113
Cantor's Pit Viper 145
Caretta caretta 26
Caucasian Agama 40
Central Asian Tortoise 15
Ceratophora aspera 34
   stoddartii 34
   tennentii 35
Cerberus rynchops 140
Chalcidoseps thwaitesii 67
Chamaeleo zeylanicus 44
Checkered Keelback Water Snake 131
Chelonia mydas 27
Chitra indica 23
Chrysopelea ornata 98
   taprobanica 98
Clerk's Keelback 125
Cnemaspis assamensis 46
   indica 47
   kandiana 47
   tropidogaster 48
   yercaudensis 48
Coelognathus flavolineatus 99
   helena 99
   radiatus 100
Collared Black-headed Snake 119
Collared Keelback 129
Common Asian Leopard Gecko 63
Common Bronzeback Tree Snake 103
Common Indian Cat Snake 97
Common Lanka Skink 73
Common Rough-sided Snake 123
Common Sand Boa 88
Common Smooth Water Snake 140
Common Vine Snake 91
Common Wolf Snake 106
Cophotis ceylanica 35
Copper-headed Trinket Snake 100
Coryphophylax subcristatus 36
Crocodylus palustris 155
   porosus 156
Crossobamon orientalis 49
Crowned River Turtle 19
Cryptelytrops albolabris 144

Index

andersoni 144
cantori 145
erythrurus 145
septentrionalis 146
Cuora amboinensis 17
mouhotii 18
Cyclemys gemeli 18
Cylindrophis maculata 83
Cyrtodactylus adleri 49
collegalensis 50
fasciolatus 50
khasiensis 51
rubidus 51
yakhuna 52
Cyrtopodion kachhense 53
Daboia russelii 146
Dark-bellied Keelback Water Snake 130
Dasia nicobarensis 68
Deignan's Lanka Skink 72
Dendrelaphis andamanensis 100
caudolineolatus 101
cyanochloris 101
humayuni 102
proarchos 102
tristis 103
Depressed Gecko 55
Deraniyagala's Lanka Skink 73
Dibamus nicobaricum 81
Dice-like Trinket Snake 92
Dog-faced Water Snake 140
Dopasia gracilis 80
Draco norvilii 36
Drummond-Hay's Shieldtail 84
Duméril's Kukri Snake 114
Dussumier's Litter Skink 78
Eastern Cat Snake 94
Eastern Fan-throated Lizard 43
Eastern Glass Snake 80
Eastern Indian Leopard Gecko 63
Eastern Keelback 127
Eastern Rat Snake 117
Eastern Trinket Snake 115
Echis carinatus 147
Elongated Tortoise 13
Emma Gray's Forest Lizard 30
Enhydris enhydris 140

Enhydris sieboldii 141
Eremias acutirostris 64
Eretmochelys imbricata 27
Eristicophis macmahoni 147
Eryx conicus 88
johnii 89
tataricus 89
whitakeri 90
Eublepharis hardwickii 63
macularius 63
Euprepiophis mandarinus 103
Eutropis andamanensis 68
carinata 69
dissimilis 69
macularia 70
multifasciata 70
quadricarinata 71
tamanna 71
tytlerii 72
False Bowring's Gecko 54
Flat-backed Mountain Lizard 38
Flat-tailed Gecko 59
Fordonia leucobalia 141
Forsten's Cat Snake 94
Four-clawed Gecko 53
Four-keeled Grass Skink 71
Four-toed Skink 67
Gans's Lanka Skink 74
Garden Lizard 33
Garnot's Gecko 57
Gavialis gangeticus 156
Gehyra mutilata 53
Gekko gecko 54
Gemel's Leaf Turtle 18
Geochelone elegans 13
Geoclemys hamiltonii 19
Gerarda prevostiana 142
Glossy Marsh Snake 142
Gloydius himalayanus 148
Gonyosoma frenatum 104
oxycephalum 104
prasinum 105
Green Cat Snake 93
Green Fan-throated Lizard 41
Green Forest Lizard 29
Green Keelback 128
Green Rat Snake 118

Green Trinket Snake 105
Green Turtle 27
Grey Kukri Snake 112
Günther's Rough-sided Snake 123
Hardella thurjii 19
Hardwicke's Spiny-tailed Lizard 41
Hawksbill Sea Turtle 27
Hebius beddomii 124
clerki 125
khasiense 125
parallelum 126
venningi 126
xenura 127
Hemidactylus aquilonius 54
brookii 55
depressus 55
flaviviridis 56
frenatus 56
garnoti 57
hunae 57
lankae 58
leschenaultii 58
parvimaculatus 59
platyurus 59
reticulatus 60
triedrus 60
Hemiphyllodactylus aurantiacus 61
typus 61
Hemprich's Shieldtail 85
Heosemys depressa 20
Herpetoreas platyceps 127
Himalayan Keelback 128
Himalayan Litter Skink 79
Himalayan Pit Viper 148
Himalayan Trinket Snake 116
Hook-nosed Sea Snake 136
Horsfield's Spiny Lizard 42
Hump-nosed Lizard 39
Hump-nosed Pit Viper 148
Hydrophis cyanocinctus 136
schistosus 136
Hypnale hypnale 148
nepa 149
Indian Black Turtle 21
Indian Chameleon 44
Indian Coral Snake 135
Indian Day Gecko 47

Index

Indian Flapshell Turtle 24
Indian Fringe-toed Lizard 64
Indian Gharial 156
Indian Golden Gecko 45
Indian Krait 133
Indian Peacock Softshell Turtle 25
Indian Rat Snake 118
Indian Rock Python 87
Indian Roofed Turtle 22
Indian Sandfish 78
Indian Softshell Turtle 24
Indian Star Tortoise 13
Indian Trinket Snake 99
Indotestudo elongata 13
   travancorica 14
Indotyphlops braminus 154
   jerdoni 154
Iridescent Snake 92
Japalura andersoniana 37
   kumaonensis 37
   planidorsata 38
Jerdon's Blind Snake 154
Jerdon's Forest Lizard 31
Jerdon's Pit Viper 150
Kandy Day Gecko 47
Kashmiri Rock Agama 38
Keeled Box Turtle 18
Keeled Grass Skink 69
Khasi Hills Bent-toed Gecko 51
Khasi Hills Keelback 125
Khasi Hills Long-tailed Lizard 66
Khasi Hills Trinket Snake 104
King Cobra 139
Kollegal Ground Gecko 50
Kumaon Mountain Lizard 37
Ladakhi Rock Skink 67
Lankascincus deignani 72
   deraniyagalae 73
   fallax 73
   gansi 74
   taylori 74
Laotian Wolf Snake 109
Large Blind Snake 153
Large-eyed False Cobra 132
Large-scaled Shieldtail 86
Large-toothed Wolf Snake 109
Laticauda colubrina 137

Laudakia tuberculata 38
Laungwala Toad-headed Lizard 29
Leaf-nosed Lizard 35
Leaf-nosed Viper 147
Lepidochelys olivacea 28
Lepidodactylus lugubris 62
Leschenaulti's Lacerta 65
Liopeltis frenatus 105
   stoliczkae 106
Lipinia macrotympana 75
Lissemys ceylonensis 23
   punctata 24
Loggerhead Sea Turtle 26
Lycodon aulicus 106
   carinatus 107
   fasciatus 107
   hypsirhinoides 108
   jara 108
   laoensis 109
   septentrionalis 109
   striatus 110
   zawi 110
Lygosoma albopunctata 75
   bowringii 76
   punctata 76
Lyriocephalus scutatus 39
   ridgewayi 111
MacClelland's Coral Snake 139
Macropisthodon plumbicolor 128
Malayan Box Turtle 17
Malayopython reticulatus 86
Mandarin Trinket Snake 103
Manouria emys 14
Many-lined Grass Skink 70
Many-spotted Cat Snake 95
Maria's Lizard 32
Medo Pit Viper 152
Melanochelys tricarinata 20
   trijuga 21
Millard's Hump-nosed Pit Viper 149
Mock Viper 117
Monocled Cobra 137
Montane Slug-eating Snake 143
Mourning Gecko 62
Moustached Forest Lizard 32
Mugger Crocodile 155
Naja kaouthia 137

   naja 138
   sagittifera 138
Narrow-headed Softshell Turtle 23
Nessia bipes 77
   monodactylus 77
Nicobar Bent-toed Gecko 49
Nicobar Bronzeback Tree Snake 102
Nicobarese Cat Snake 97
Nicobarese Tree Skink 68
Nicobarese Worm Lizard 81
Nilssonia gangeticus 24
   hurum 25
   nigricans 25
North-eastern Water Skink 80
Northern Pit Viper 146
Norville's Flying Lizard 36
Occelated Ground Agama 44
Oceanic Worm Gecko 61
Oligodon albocinctus 111
   arnensis 112
   cinereus 112
   cyclurus 113
   dorsalis 113
   sublineatus 114
   taeniolatus 114
Olive Keelback Water Snake 124
Olive Ridley Sea Turtle 28
Ophiomorus raithmai 78
Ophiophagus hannah 139
Ophisops jerdoni 65
   leschenaultii 65
Oreocryptophis porphyraceus 115
Oriental Vine Snake 91
Ornate Flying Snake 98
Orthriophis cantoris 115
   hodgsonii 116
   taeniurus 116
   wiegmanni 39
Ovophis monticola 149
Painted Bronzeback Tree Snake 102
Painted Roofed Turtle 17
Painted-lipped Lizard 30
Pakistani Ribbon Snake 121
Pangshura smithii 21
   sylhetensis 22
   tectum 22
Paralaudakia caucasia 40

Index

Pareas macularius 142
   monticolus 143
Pelochelys cantorii 26
Phelsuma andamanense 62
Popeia popeiorum 150
Pope's Pit Viper 150
Protobothrops jerdonii 150
   mucrosquamatus 151
Psammodynastes pulverulentus 117
Psammophilus dorsalis 40
Psammophis leithii 121
Pseudoxenodon macrops 132
Ptyas korros 117
   mucosa 118
   nigromarginata 118
Ptyctolaemus gularis 41
Pygmy Lizard 35
Python bivittatus 87
   molurus 87
Red Sand Boa 89
Red-necked Keelback 129
Red-tailed Pit Viper 145
Red-tailed Trinket Snake 104
Reticulated Gecko 60
Reticulated Python 86
Rhabdophis himalayanus 128
   nuchalis 129
   subminiatus 129
Rhabdops bicolor 119
Rhinoceros-horned Lizard 34
Rhinophis blythii 84
   drummondhayi 84
   homolepis 85
River Terrapin 16
Rough-bellied Day Gecko 48
Rough-horned Lizard 34
Russell's Viper 146
Saara hardwickii 41
Salea horsfieldii 42
Saltwater Crocodile 156
Saw-scaled Viper 147
Sharp-nosed Racerunner 64
Sibynophis collaris 119
   sagittarius 120
   subpunctatus 120
Siebold's Water Snake 141
Sindh Sand Gecko 49

Sinomicrurus macclellandi 139
Sitana bahiri 42
   ponticeriana 43
Small-eared Striped Skink 75
Smith's Snake Skink 77
Snake-eyed Lacerta 65
South Indian Rock Agama 40
Spectacled Cobra 138
Sphenomorphus dussumieri 78
   indicus 79
   maculatus 79
Spot-tailed Kukri Snake 113
Spotted Litter Skink 79
Spotted Pond Turtle 19
Spotted Slug-eating Snake 142
Spotted Supple Skink 76
Sri Lanka Wolf Snake 107
Sri Lankan Cat Snake 93
Sri Lankan Fan-throated Lizard 42
Sri Lankan Flapshell Turtle 23
Sri Lankan Flying Snake 98
Sri Lankan Golden Gecko 46
Sri Lankan Green Pit Viper 152
Sri Lankan Kangaroo Lizard 39
Sri Lankan Keelback Water Snake 130
Sri Lankan Krait 134
Sri Lankan Pipe Snake 83
Sri Lankan Spotted Gecko 59
Sri Lankan Spotted Rock Gecko 57
Sri Lankan Termite Hill Gecko 58
Stejneger's Pit Viper 153
Stoliczka's Stripe-necked Snake 106
Strange-tailed Keelback 127
Stripe-necked Snake 105
Stripe-tailed Bronzeback Tree Snake 101
Striped Grass Skink 69
Striped Trinket Snake 116
Sunbeam Snake 88
Takydromus khasiensis 66
Tammenna Skink 71
Tartary Sand Boa 89
Tawny Cat Snake 95
Taylor's Lanka Skink 74
Termite-hill Gecko 60
Testudo horsfieldii 15

Thai Cat Snake 96
Three-striped Roofed Turtle 16
Toeless Snake Skink 77
Tokay Gecko 54
Trachischium monticola 121
Trapelus agilis 43
   megalonyx 44
Travancore Tortoise 14
Triangled Keelback Water Snake 131
Tricarinate Hill Turtle 20
Trimeresurus gramineus 151
   trigonocephalus 152
Tropidophorus assamensis 80
Two-coloured Forest Snake 119
Tytler's Grass Skink 72
Uropeltis liura 85
   macrolepis 86
Varanus bengalensis 81
   flavescens 82
   salvator 82
Venning's Keelback 126
Viridovipera medoensis 152
   stejnegeri 153
Wart Snake 83
Warty Rock Gecko 53
Water Monitor 82
Western Ghats Worm Gecko 61
Whistling Lizard 31
Whitaker's Sand Boa 90
White-barred Kukri Snake 111
White-bellied Mangrove Snake 141
White-lipped Pit Viper 144
White-spotted Supple Skink 75
Xenochrophis asperrimus 130
   cerasogaster 130
   piscator 131
   tytleri 132
   trianguligerus 131
Xenopeltis unicolor 88
Yellow Monitor 82
Yellow-green House Gecko 56
Yellow-lipped Sea Krait 137
Yellow-speckled Wolf Snake 108
Yellow-striped Trinket Snake 99
Yercaud Day Gecko 48
Zaw's Wolf Snake 110